THE FOCAL EASY GUIDE TO
FINAL CUT PRO 4

The Focal Easy Guide Series

Focal Easy Guides are the best choice to get you started with new software, whatever your level. Refreshingly simple, they do *not* attempt to cover everything, focusing solely on the essentials needed to get immediate results.

Ideal if you need to learn a new software package quickly, the Focal Easy Guides offer an effective, time-saving introduction to the key tools, not hundreds of pages of confusing reference material. The emphasis is on quickly getting to grips with the software in a practical and accessible way to achieve professional results.

Highly illustrated in color, explanations are short and to the point. Written by professionals in a user-friendly style, the guides assume some computer knowledge and an understanding of the general concepts in the area covered, ensuring they aren't patronizing!

Series editor: Rick Young (www.digitalproduction.net)

Director and Founding Member of the UK Final Cut User Group, Apple Solutions Expert and freelance television director/editor, Rick has worked for the BBC, Sky, ITN, CNBC and Reuters. Also a Final Cut Pro Consultant and author of the best-selling *The Easy Guide to Final Cut Pro*.

Titles in the series:

The Easy Guide to Final Cut Pro 3, Rick Young

The Focal Easy Guide to Final Cut Pro 4, Rick Young

The Focal Easy Guide to Final Cut Express, Rick Young

THE FOCAL EASY GUIDE TO
FINAL CUT PRO 4

For new users and professionals

RICK YOUNG

AMSTERDAM • BOSTON • HEIDELBERG • LONDON • NEW YORK • OXFORD
PARIS • SAN DIEGO • SAN FRANCISCO • SINGAPORE • SYDNEY • TOKYO
Focal Press is an imprint of Elsevier

Focal Press
An imprint of Elsevier
Linacre House, Jordan Hill, Oxford OX2 8DP
200 Wheeler Road, Burlington MA 01803

First published 2004

Copyright © 2004, Rick Young. All rights reserved
Photographs copyright © 2004, Rick Young. All rights reserved

The right of Rick Young to be identified as the author of this work
has been asserted in accordance with the Copyright, Designs and
Patents Act 1988

No part of this publication may be reproduced in any material form (including
photocopying or storing in any medium by electronic means and whether
or not transiently or incidentally to some other use of this publication) without
the written permission of the copyright holder except in accordance with the
provisions of the Copyright, Designs and Patents Act 1988 or under the terms of
a licence issued by the Copyright Licensing Agency Ltd, 90 Tottenham Court Road,
London, England W1T 4LP. Applications for the copyright holder's written
permission to reproduce any part of this publication should be addressed
to the publisher

Permissions may be sought directly from Elsevier's Science and Technology Rights
Department in Oxford, UK: phone: (+44) (0) 1865 843830; fax: (+44) (0) 1865 853333;
e-mail: permissions@elsevier.co.uk. You may also complete your request on-line via the
Elsevier homepage (www.elsevier.com), by selecting 'Customer Support'
and then 'Obtaining Permissions'

British Library Cataloguing in Publication Data
Young, Rick
   The Focal easy guide to Final Cut Pro 4
   1. Final Cut (Computer file)   2. Digital video – Editing – Data processing
   3. Video tapes – Editing – Data processing   I. Title
   778.5'93

Library of Congress Cataloguing in Publication Data
A catalogue record for this book is available from the Library of Congress

ISBN   0 240 51925 6

For information on all Focal Press publications visit our website at:
www.focalpress.com

Typeset by Newgen Imaging Systems (P) Ltd., Chennai, India
Printed and bound in Italy

Contents

Preface xi
Introduction xiii

Getting Started 1
The Digital Laboratory 2
Hardware and Software Requirements 4
Inside your Mac 5
How Much Hard Drive Space? 5
Firewire 6
Before and After Firewire 7
Video Formats 9
Television Aspect Ratio 9
Loading the Software 10
Initial Setup 11
Easy Setup 12
DV Audio 13
Creating a DV 32 kHz Easy Setup 14

The Interface 19
Arranging the Interface 21
Learning a Custom Layout 23
Important Details about the Interface 24
Button Bars 26
Customizing the Keyboard Layout 27

Capture 29
Setting Scratch Disks 30
Working with Formats other than DV 32

Methods of Capturing DV Footage   33
Deck Control   34
The Capture Window   34
Capture Clip   36
Capture Now   37
Batch Capture   38
Selectively Capturing using Batch Capture   40
Getting the Most Out of the Capture Process   42
Importing Music from CD   42
Converting Audio Sample Rates   43

Organizing your Footage   45
Viewing Clips   46
Playing Video through Firewire   47
DV Start/Stop Detection   47
Working with Bins   49
Working in Icon View   51
Setting Poster Frames   52
Searching for Clips   53

Editing   55
Insert and Overwrite Editing   57
Getting Started with Editing   57
Distinguishing between Insert/Overwrite   61
Three Point Editing   64
Other Editing Options   65
Modifying 'In' and 'Out' Points   66
Directing the Flow of Audio/Video   67
Locking Tracks   69
Adding and Deleting Tracks   70
Essential Editing Tools   71

Undo/Redo 72
Linked/Unlinked Selection 73
Moving Edits in the Timeline 75
Selecting Multiple Items in the Timeline 76
Cut, Copy, Paste 78
Snapping and Skipping between Shots 79
The Razorblade Tool 79
The Magnifier Tool 81
Bringing Clips Back into Sync 82
Creating New Sequences 83
Subclips 84
Freeze Frame 85
Match Frame Editing 86
Slow/Fast Motion 88
Split Edits 90
Drag and Drop Editing 93
Extending/Reducing Clips by Dragging 96

Rendering 99
The Render Settings 101

Media Management 107
Making Clips Offline 108
The Render Manager 109

Effects 113
The Concept of Media Limit (Handles) 117
Applying Transitions 117
Changing Transition Durations 118
Applying Filters 119
Compositing 122

Methods of Creating Multiple Tracks   123
The Motion Tab   124
Using the Motion Tab   124
Image + Wireframe   128
Titlesafe   130
Working with Multi-Layers   131
Keyframing Images   133
Multi-Layered Dissolves   141
Keyframing Filters   143
Time Remapping   146
Copy and Pasting Attributes   151
Titling   152

Audio   159

Setting Correct Audio Levels   160
Getting the Most Out of your Audio   161
Converting Clips into Stereo Pairs   161
Adjusting Audio Levels   163
Adding Sound Fades   164
The Audio Mixer   166
Adjusting and Recording Audio Keyframes   168
Adding Audio Cross Fades   169
Adding Audio Tracks   170
Mixdown Audio   171

Output   173

Print to Video   174
Other Forms of Distribution   176
Export Using Compressor   177
Export Using QuickTime Conversion   180

LiveType 183

LiveType Templates 185
Replacing Text in the Templates 187
Replacing Video in the Templates 188
Working Inside LiveType 189
Media Browser 190
Creating Results Manually in LiveType 193
Output from LiveType 196

Soundtrack 199

IMPORTANT: before you begin 200
Exporting for Soundtrack from Final Cut Pro 202
The Soundtrack Workflow 204
Importing a Movie into Soundtrack 205
Building your Mix 206
Exporting the Mix 210

Epilog 212

Index 215

Preface

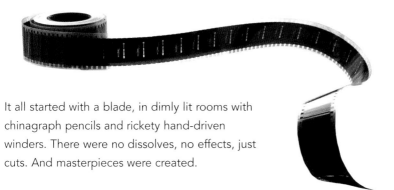

It all started with a blade, in dimly lit rooms with chinagraph pencils and rickety hand-driven winders. There were no dissolves, no effects, just cuts. And masterpieces were created.

Chaplin, Eisenstein, Welles, Capra, Wilder, Hitchcock – each and every one of these great directors worked with far less sophisticated equipment than you and I, and, yet, truly great films were made.

It may have all started with a blade, yet in the modern world editing takes place on sophisticated work stations with more power than the computers used to send men to the moon! And the cost of these work stations has plummeted to an all time low. With nothing more than a DV camera, Firewire Mac and Final Cut Pro, one has tools which are infinitely superior to filmmakers of previous generations.

Yet, no matter how great the tools, it is essential to understand the principles of filmmaking. The limitations are no longer access to equipment or technology. It all comes down to skill and craftsmanship.

Introduction

When Final Cut Pro was first released in the last year of the twentieth century the world of video acquisition and post-production was going through a period of upheaval. High quality, low-cost DV cameras had made significant in-roads into both professional and consumer markets and the world of editing was undergoing a similar transformation.

Initially, Final Cut Pro was perceived as a DV editor, however, early pioneers of the digital video revolution soon put the application to test in the real world of professional post-production. There were those who invested in high-end video cards, editing everything from uncompressed video to professional DV, while others used the program to explore its potential for compositing and effects work. Final Cut Pro, in combination with other applications, soon infiltrated every area of the production scene, being used for network broadcast, feature film production, Internet, CD and DVD production.

With the introduction of Final Cut Pro 4 it is no longer fair to call this a single application – rather Final Cut Pro is now a suite of five applications. There is Final Cut Pro – for editing and compositing work; LiveType – for the creation of complex moving titles; Soundtrack – for creating musical scores; Compressor – for exporting files for DVD, CD and Internet preparation; and Cinema Tools – for film production.

Most of this book focuses on getting to grips with Final Cut Pro as an editing application. The new programs, Compressor, Soundtrack and LiveType, are all dealt with towards the end of the book.

So get ready for a ride – one that will take you through all the processes needed to structure, edit and output a film. The power to create, like never before, is at your fingertips.

Rick Young
Producer/Director/Editor
London, UK

Thanks to everyone in the Final Cut Pro community.
You know who you are.

GETTING STARTED

The Digital Laboratory

Think of your computer loaded with Final Cut Pro as being like a digital laboratory. In the days when cine-film was the only means for movie-making, everyone relied on the lab. Film would be processed at the lab; there were work prints; answer prints; release prints; opticals . . . the lab was central to virtually every facet of the post-production process.

Your Mac is a digital lab, just waiting for you to stir the potions.

Essentially the post-production process is the same as it has always been. While the means to achieving results has changed, digital filmmaking requires similar methods and procedures to that of filmmaking in the world of celluloid and chemicals.

While film needed to be developed the images recorded on videotape need to be transferred from tape to hard drive – this process is known as **capture**.

The raw material must then be ordered and structured. In the film world this would take place in the cutting room where the editor would take reels of film and break these into smaller more manageable sections – when using Final Cut Pro an electronic equivalent to the cutting room is provided in the layout of the interface. It is here that the **editing** takes place.

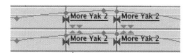

Once the picture was edited the sound must be **mixed**. Dubbing suites with many machines running in synchronization were traditionally used. Inside your computer multiple audio tracks are electronically mixed to be output in sync with picture.

Music for film in the early days of film production was recorded by a team of session musicians or even an entire orchestra. Later musical scores were accessed from stock or production music libraries. Music can now be created using **Soundtrack**, a stand-alone application provided with Final Cut Pro.

Effects and titles were traditionally created using a device known as an optical printer. Film exposed in the optical printer would then be immersed in developing tanks, in total darkness, to emerge, as if by magic, with hundreds of tiny transparent images. When projected these images would light up a room . . . Final Cut Pro uses electronic processes to achieve these results. Video tracks are layered in order of priority to build effects which can be made up of many different layers. This process is known as **compositing**.

Complex titles were produced in optical houses along with the other effects. **LiveType** is an application specifically designed for title creation and is fully capable of making titles comparable to those seen on network television.

Finally, the original negative would be cut and matched by technicians, wearing pure white gloves, in dust-proof rooms. Release prints were produced so the

film could be distributed to cinemas and later television stations throughout the world. Release prints in the modern world are recorded onto digital tape, DVD or the final edit may need to be prepared for CD-ROM or Internet delivery. This phase of the process is known as **output**.

It should be obvious that a distinct set of processes takes place in the editing of any production. When using Final Cut Pro these processes can be broken down into five distinct areas.

(i) capture (ii) editing (iii) sound mixing (iv) compositing (v) output

Learn how to perform these essential tasks and you will be well armed with the knowledge needed to edit any program. Once these processes are learnt, you, as the editor, will be able to concentrate on the creative aspects of the editing process. Only when one moves beyond the mechanics of the editing can Final Cut Pro be used to its full potential.

Hardware and Software Requirements

The hardware you need to use Final Cut Pro:

1   A Firewire camera or DV deck.

2   A G4 or G5 Mac with at least 384 MB of ram and enough free hard drive space to store your video files. To access the Realtime effects you will need a G4 processor running at 500 MHz or above with 512 MB of ram. A total of 15 GB is required for a full installation of Final Cut Pro 4 and the content supplied with LiveType and Soundtrack.

3   OS 10.2.5 or greater and QuickTime 6.1 or greater.

Your digital video deck or camera will need DV 'in' and 'out'. Ideally you should also have a television monitor and a pair of external speakers.

None of the G3 machines work with Final Cut Pro 4.

It should be noted that Final Cut Pro 4 can be installed on its own, thus Soundtrack and LiveType do not need to be installed to use Final Cut Pro as an editing application only. This is a big advantage if you are limited with hard drive space.

Inside your Mac

Your Mac is made up of many different components all specifically engineered to work together. There are hard drives, fans, a motherboard, memory, circuits, a power supply, ports and slots. Data pumps through the internal system while the keyboard and mouse act as the interface between the computer and the mind of the operator. While it is not essential to understand exactly what goes on inside your Mac it is helpful to have a general overview – particularly with regards to memory and available hard drive space. These two areas are critical to having an efficient and well-managed machine.

How Much Hard Drive Space?

The hard drives are the place where you store your video files. Any video editing system requires large hard drives capable of storing vast amounts of data. While 'the more, the better' rule applies, each and every one of us is on some sort of limit and we all have to stop somewhere. I would suggest a system drive above 30 GB as this will allow room for a full installation of Final Cut Pro plus the 14 GB of content supplied with Soundtrack and LiveType. In addition, a second hard drive of 60 GB drive or more is desirable.

A lot has been written over the years of the benefits of working with the operating system of your computer on one drive and storing your captured clips to a separate drive. This is really the best way to configure your system, but in the real world, a lot of people will have to use a single hard drive for the operating system and media storage for the simple reason that they only have one hard drive physically installed inside their computer. The professional Macs allow for additional drives to be installed in the computer – G4 towers have room for four internal drives, while the G5s have room for two internal drives.

Should you require more drives than your computer allows the simplest option is to go for external Firewire drives.

Video at DV resolution chews up approximately 1 GB to 4.5 minutes of sound and video. It is easy, therefore, to work out how much material you can store on hard disk. Simply multiply the capacity of your hard drive by 4.5 and then divide the result by 60. This will calculate the amount of storage you will get in hours and minutes. The measurement of 4.5 minutes to the gigabyte is a conservative estimate. You actually get slightly more. Therefore a 60 GB drive will provide room for between 4 and 5 hours of digital video. A 200 GB drive stores approximately 15 hours at DV resolution.

If you are working with formats of higher resolution than DV then the amount of storage per gigabyte drops dramatically. While one minute of DV footage consumes 216 MB, one minute at the uncompressed standard will use 1.4 GB, and at high definition the same one minute will eat up 7.3 GB of hard drive space. Furthermore, uncompressed and high definition video formats are far more demanding and often require expensive RAIDs and SCSI drives which spin much faster than standard IDE drives.

Firewire

Firewire is an Apple invented technology which also goes under the name of iLink and IEEE1394. One of the remarkable features of Firewire technology, unlike USB, is that it is intelligent. Firewire serves not only as a data transfer bus, it also allows for device control. It is for this reason that video and audio can be transferred through a Firewire cable and deck control can take place. Furthermore, Firewire is also bi-directional which means video and audio can flow in both ways through the cable. Thus video and audio can be transferred from a deck/camera to a computer and then back again – or, alternatively, one can perform deck to deck editing.

The golden rule when connecting your Mac and camera/deck with a Firewire cable is to make sure the connector is the correct size. Without sounding too basic, make sure you insert the Firewire connector correctly – if you jam it in backwards you will end up with a bent Firewire port.

Firewire ports are identified by a symbol (which looks remarkably similar to a nuclear warning symbol).

Small and Large Firewire Connectors   Firewire 400   Firewire 800

Firewire cables come in several forms. Cables can be made up of any combination of small to small, large to large, or small to large connectors. The larger 6-pin Firewire port is found on the back or side of your Mac (depending which Mac you have), while the smaller 4-pin Firewire connector is located on your camera or deck. The latest version of Firewire, known as Firewire 2, or IEEE 1394b, has a maximum transfer speed of 800 Mb per second which is twice the speed of the original version, known as Firewire 400.

Simply plug the large end of the Firewire cable into your Mac and the small end into your camera or deck. Firewire cables are hot-pluggable which means they can be connected or disconnected while the Mac is switched on or off, although, ideally, the devices should be plugged together prior to launching Final Cut Pro. Otherwise a warning message will appear to alert you to the fact that no Firewire device in being seen.

Before and After Firewire

It was all analog. Everything tangled up in a mass of cables. There were wires everywhere and different standards too. We're talking 1980s' technology. Composite Video, S-Video, Component Video.

All through the 1980s the standard was Sony Betacam. First there was standard Betacam, followed by Betacam SP and eventually, well into the 1990s, Digibeta emerged as the standard for professional digital production. Sony may have lost the format war to VHS but when it came to the professional arena Sony was untouchable.

Before Betacam it was U-matic, available in low-band and hi-band versions. There were various one inch formats: A, B and C. C-format was the best by a long shot. It was like working with 35 mm film and coincidentally the tape was about the same in measurement. Before one inch there was Quad – two-inch tape that originated in the 1950s when Ampex first invented videotape.

The 1990s. Digital is everywhere. Digibeta, D1, D2, D3, D5, D9. Avid ruled the non-linear market, with Media 100 chasing at its heels. DV hadn't even been invented. Final Cut Pro wasn't even a whisper.

Everything changed in 1996 with the introduction of one camera: the Sony VX-1000. When this camera appeared on the market the world went crazy. I remember the BBC had purchased 100 of these and the camera had only just been released.  Then I started hearing the BBC had a VX-1000 in every single department in the whole of the BBC. Documentaries were filmed with this camera, multi-camera shoots were produced and the professional world with all their big cameras sat back in astonishment as the world of acquisition was redefined, apparently, overnight!

DV blew the whole scene apart. The quality of DV, as a recording format, is equivalent to Beta SP. Perhaps on a technical chart DV might score slightly less, but then DV doesn't suffer from the drop-out problem which plagued Beta SP due to shedding and flaking of oxide.

DV was just the sort of technology the world wanted desperately. Finally, a low-cost, lossless, high quality camera/editing solution had arrived. This exact same technology forms the basis of Firewire editing systems today – only the deck or camera is usually connected to a computer rather than editing from camera to deck or vice versa.

When using a Firewire-based editing system you can work with either a camera or deck – providing the camera has both Firewire 'in' and 'out'. The advantages of having a deck are (i) you don't beat up your camera every time you capture footage (ii) a deck offers other features such as different inputs, the ability to work with large or small size tapes, a large time-code display, a jog/shuttle wheel, and often a built-in edit controller.

Video Formats

The world we live in operates with several different video formats.

DV-NTSC applies to the USA, Japan and many other parts of the world; whereas DV-PAL is used throughout most of Europe, Australia, and parts of Asia. There are other formats such as SECAM which is used in France and North Africa, and variations on PAL and NTSC are used in South America. However, PAL and NTSC remain the dominant formats.

It is simple but crucial to set the correct video format when working with Final Cut Pro.

Final Cut Pro 4 is geared to edit everything from DV to standard and high definition video. It is also capable of film production.

Television Aspect Ratio

Another consideration is whether the footage you are working with has been filmed in widescreen anamorphic – 16 × 9, or standard television format – 4 × 3. Do not confuse letterbox (cropped 4 × 3) with true widescreen. Most consumer cameras do not offer a true widescreen anamorphic mode of operation. However, many offer a cropped 4 × 3 letterbox setting.

Standard 4 × 3
Television Format

Widescreen
Anamorphic

Cropped 4 × 3
or Letterbox

You will need to consult your camera manual to determine the settings available for the camera you are using.

Loading the Software

1 Put the Final Cut Pro 4 DVD into the DVD drive of your Mac.

2 Double click the Install Final Cut Pro 4 icon.

Install Final Cut Pro 4

Follow the on-screen instructions, making sure you agree to the licence agreement.

3 You will need to enter details including the serial number prior to installation. Once you have the correct serial number entered press OK.

4 Highlight the drive you wish to install the software onto and click continue. The installation process will now commence.

Preparing Final Cut Pro 4
Configuring Installation

Rick's Hard Disc
34.9GB (30.8GB Free)

5 Once the software is installed you will be prompted to restart your Mac.

After you have restarted your machine a message will prompt you to run the installers for LiveType and Soundtrack. If you choose you can begin using Final Cut Pro straightaway. However, if you wish to install the other applications – LiveType and Soundtrack – then you need to load the data from the other DVDs that are included in the box.

The order in which you install LiveType and Soundtrack doesn't matter. Make sure, however, that you install the LiveType data using disc one followed by disc two.

The process for installing LiveType and Soundtrack is virtually identical to the installation process you have just gone through when installing Final Cut Pro. The complete installation process takes about an hour. Once it is complete you need to update your QuickTime components to QuickTime Pro.

1 Select the Apple menu at the top left of the screen and scroll to System Preferences.

Finder   File   Edit
About This Mac
Get Mac OS X Software..

System Preferences...

GETTING STARTED

2 Click on the QuickTime icon.

3 Click Registration.

4 Complete the details. Make sure you enter QuickTime where it says Registered to – with a capital Q and T. Enter the QuickTime Pro serial number, making sure you do not confuse the letter 'o' with the number 0.

5 Click OK.

Initial Setup

Once Final Cut Pro has been successfully installed you need to be able to access Final Cut Pro and the other applications which are installed on your system.

Final Cut Pro, LiveType, Soundtrack, Compressor and Cinema Tools are all found inside the Applications folder which is located on the hard drive where the operating system for your computer is installed.

The easiest way to get to the Applications folder is to choose the menu at the top of the desktop screen titled Go.

1 Select the Go menu and scroll down to Applications.

2 Locate the Final Cut Pro application icon in the Applications folder.

3 Drag the Final Cut Pro icon onto the dock.

11

4   Do the same with LiveType, Soundtrack, Compressor and Cinema Tools.

5   Click once on the Final Cut Pro icon to launch the program. You are now ready to begin work within Final Cut Pro.

Easy Setup

Apple have made it very easy to set up Final Cut Pro, however, it is up to you to make sure you get these settings right. If you set the audio sample rate incorrectly the result will be sync drift; if you set the video to widescreen anamorphic when it was shot in standard 4 × 3 your images will not fit the frame correctly.

The simplest way to set up Final Cut Pro is to access the Easy Setup menu.

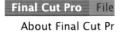

1   Open the Final Cut Pro menu at the top left of the screen. Scroll to Easy Setup and release your mouse button.

2   Click with your mouse on the setup bar within Easy Setup and choose the relevant option.

It is possible to extend the range of Easy Setups by clicking the Show All box to the right of the Easy Setup bar.

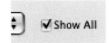

GETTING STARTED

Cinema Tools - 24fps from DV PAL
Cinema Tools - DV NTSC NDF
DV-NTSC
DV-NTSC 24p (23.98)
DV-NTSC 24p (23.98) Advanced Pulldov
DV-NTSC Anamorphic
DV-NTSC FireWire Basic
DV-PAL
DV-PAL Anamorphic
DV-PAL FireWire Basic
DV50 - NTSC
DV50 - PAL
DV50 NTSC 24p (23.98)
DV50 NTSC Anamorphic
DV50 PAL Anamorphic
DVCPRO - PAL
DVCPRO - PAL 48 kHz Anamorphic
OfflineRT - DVCPRO - PAL
OfflineRT HD - 23.98
OfflineRT HD - 24
OfflineRT HD - 25
OfflineRT HD - 29.97
OfflineRT HD - 30
OfflineRT NTSC
OfflineRT NTSC 24fps
OfflineRT NTSC 24p (23.98)
OfflineRT NTSC Anamorphic
OfflineRT PAL
OfflineRT PAL Anamorphic
Uncompressed 10-bit NTSC 48 kHz
Uncompressed 10-bit PAL 48 kHz
Uncompressed 8-bit NTSC 48 kHz
Uncompressed 8-bit PAL 48 kHz

The range of Easy Setups is extensive, with options for Widescreen Anamorphic, DVC Pro 50, and Uncompressed video.

There are also settings for working with FireWire Basic. The FireWire Basic setting has been included because different manufacturers use slightly different ways of implementing the Firewire standard. If you do encounter problems controlling your Firewire device then it is worth trying one of the FireWire Basic settings.

It is important to be aware that the DV-PAL and DV-NTSC Easy Setups both have audio sample rates set to 48 kHz. Should you be working with audio recorded at 32 kHz then you will need to create a new Easy Setup and set the sample rate accordingly.

A detailed description of how to create a 32 kHz Easy Setup is explained in the following sections.

DV Audio

It is important to understand how DV audio works. Otherwise you can end up in a lot of trouble when you begin the capture process. DV audio can be recorded at two different sample rates:

16 bit – 48 kHz
12 bit – 32 kHz

13

16 bit – 48 kHz provides the highest quality and allows for two channels of audio, or a single stereo pair to be recorded.

12 bit – 32 kHz provides lesser quality, though still very good, and allows for two sets of stereo pairs, or four individual tracks to be recorded.

By default, the DV-PAL and DV-NTSC Easy Setups are set to the highest possible sample rate of 48 kHz.

For the most, 16 bit audio is the preferred option, unless one specifically needs to access four independent channels. While this may sound ideal no DV cameras actually have inputs to record four independent channels of audio. The main advantage to setting a camera to 12 bit – 32 kHz is that audio dubbing can then take place onto the remaining free set of stereo pairs.

Unless you specifically plan on accessing these tracks I recommend setting your camera to 16 bit – 48 kHz. This is usually accessed through the menu settings in your camera.

The Audio Sample Rate is Set in the Menu of your DV Camera

Creating a DV 32 kHz Easy Setup

As described, when working with footage acquired on DV it is essential to get the audio sample rate right. Working with an incorrect sample rate in relation to the settings in Final Cut Pro is the most common cause of sync drift.

While the supplied Easy Setups cover most editing requirements, there are some situations where you may find it necessary to create additional Easy Setups; for example if you wish to work with DV audio set to the sample rate of 32 kHz. To achieve this you will need to first create a DV 32 kHz Capture Preset, in either PAL or NTSC, and then save this as an Easy Setup.

1 Select the Final Cut Pro menu.

GETTING STARTED

2   Scroll to Audio/Video Settings.

3   Choose the third tab to the right titled Capture Presets.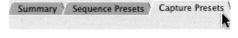

4   Choose the video format of your choice, such as DV PAL 48 kHz or DV NTSC 48 kHz.

    DV NTSC 48 kHz Anamorphic
    ✓ DV PAL 48 kHz
    DV PAL 48 kHz Anamorphic
    DV to OfflineRT NTSC (Photo JPEG)

5   Click Duplicate.

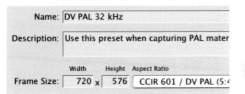

6   Name the setup, for example DV PAL 32 kHz.

7   Click the box labelled Rate and change the Audio Sample Rate to 32 kHz.

8   Click OK.

The new setting will now appear in the list of presets.

Now that you have created a DV 32 kHz Capture Preset it is necessary to create and save this as an Easy Setup.

1   Click the Final Cut Pro menu at the top left of the screen.

2   Scroll to Audio Video Settings and release your mouse button.

DV NTSC 48 kHz Adva
DV NTSC 48 kHz Anan
DV PAL 32kHz
DV PAL 48 kHz
DV PAL 48 kHz Anamc

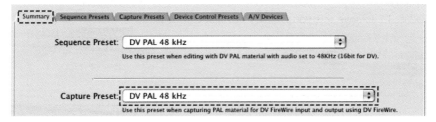

15

3   Immediately in front of you, under the Summary tab, click on Capture Preset which will reveal a list of options.

4   Choose the DV 32 kHz setting which you just created.

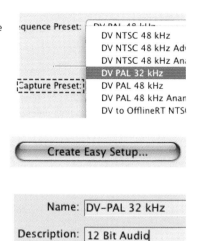

5   Still under the Summary tab, choose Create Easy Setup located at the bottom of the window.

6   A dialog box will appear. Enter the name of the Easy Setup you wish to create.

7   Enter a description for the setup.

8   Save the setup in the Custom Settings area and press OK.

This new setup will now be accessible from the Easy Setup menu.

The above procedure can be followed to create any Easy Setup you may require.

Note: Once an Easy Setup has been chosen you can confirm the setting by opening the Log and Capture window.

1   Choose the File menu towards the top left of the screen.

2   Scroll to Log and Capture and release your mouse button.

3   The Log and Capture window will open.

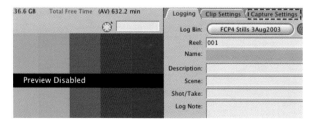

16

4   Choose Capture Settings, which is the third tab to the right.

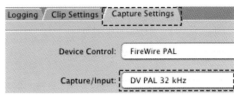

5   Look to the Capture/Input label below the Capture Settings tab in the Log and Capture window.

6   The Easy Setup you have chosen will be displayed.

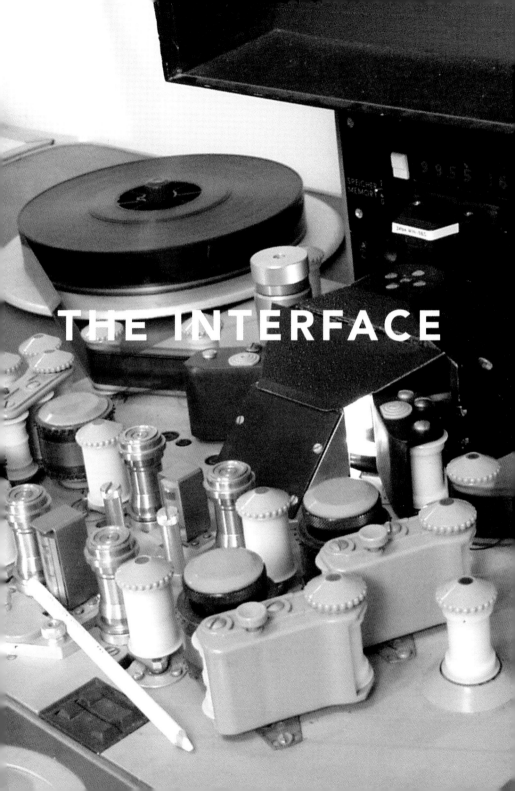

THE INTERFACE

> I fell in love with Final Cut Pro the first time I saw it and knew that it was going to be an app. that changed the world.
> GARY ADCOCK
> CHICAGO FINAL CUT PRO USER GROUP

As much of the post-production industry moved from film to video production a new way of working came into being. Moviolas and flat-beds had competition to deal with as a new kid appeared on the block. The kid was called the two-machine video editing suite.

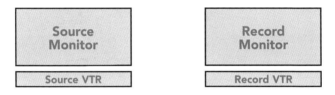

In this environment the editor would line up a shot on a source machine and edit across to a record machine. 'In' and 'out' points were marked, tapes pre-rolled, then run up to speed and images in the form of electronic signals were copied from one machine to the other.

The Final Cut Pro interface is modelled on the same idea.

The Interface is Made up of Four Main Windows

THE INTERFACE

Notice the two windows located at the top right of the Final Cut Pro interface. Think of the window on the left, the Viewer, as being the source monitor and the window on the right, the Canvas, as being the record monitor. In essence, a shot is lined up in the Viewer and copied across to the Canvas. The area immediately below the Viewer and Canvas is known as the Timeline. This shows the edited shots as blocks in the order in which they have been edited. The left window above the Timeline is called the Browser. Think of this as being like a cabinet which stores the masses of footage ready for the editor to access.

Also, take note of the Audio Meters and Tool Palette.

All professional VTRs have meters which must be watched to make sure the audio doesn't distort during the transfer and playback of sound and picture. The golden rule is always to make sure the audio meters do not peak into the red (DV audio should peak somewhere between −12 dB and −6 dB).

To swing the analogy back to the film days the Tool Palette represents the tools the editor would physically work with: the splicer, the hand-winders, the spools and frame measuring instruments. The Tool Palette in Final Cut Pro gives the editor access to the instruments with which the finer details of the editing process are crafted. It should be clear by now that Final Cut Pro draws on the very best the world of post-production has offered in the history of film and video production.

If you find the terms Browser, Viewer, Canvas, and Timeline difficult to identify with, just think of the Browser as the place where all clips are stored, the Viewer is where one watches the individual video clips, the Canvas is where the material is edited and the Timeline is the place where the individual shots which make up the entire movie are arranged.

Arranging the Interface

The Final Cut Pro interface can be set up in several different ways. Individual users can work according to their own particular preference. Several different arrangements can be chosen from within Final Cut Pro or the editor can create their own custom layouts.

1 Go to the Window menu (located top right) and scroll down to Arrange. You will notice there is a list of options for arranging the interface.

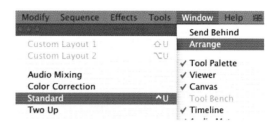

2 Scroll to any of these options and release your mouse button. Each time you wish to try a different layout you need to return to the Window menu, scroll to Arrange and then move across to the layout you wish to select.

My preference is to use the Standard setting, however, I do modify this setup slightly.

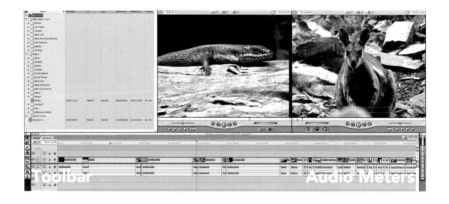

I position the Toolbar to the left, the Timeline in the center and the Audio Meters to the right. This produces a neat, symmetrical display.

To achieve this setup is simple:

1 Drag the Toolbar by clicking in the gray area at the top and position it on the opposite side of the screen beneath the Browser and next to the Timeline.

2 Slide the Timeline to the right so that it is positioned directly between the Toolbar and the Audio Meters.

3 If necessary resize the Timeline window by dragging the bottom right corner so there is no overlap onto either the Toolbar or Audio Meters.

Once you have set the layout according to your personal preference it is then possible to save the setup as a Custom Layout.

Learning a Custom Layout

To set a Custom Layout, so that it can be recalled at any time, is easy to achieve. Press the key/mouse combination in the following order:

1 Hold down the Alt/Option key (located to the left of the Space Bar).

2 While still holding down the Alt/Option key, select the Window menu at the top of the screen.

3 Scroll to Arrange. Where it normally displays Custom Layout 1 it will now read Set Custom Layout 1. Point your cursor to this setting and release the mouse button. Your Custom Layout will now be set.

You can confirm this has been achieved by selecting any of the other layouts. Now go back to the Window menu, scroll to Arrange and select Custom Layout 1. Your screen should revert back to the Custom Layout you have just set. If it does not, backtrack using the instructions above and try again. Once your Custom Layout has been set it will be remembered each time you open up Final Cut Pro and you can then choose your Custom Layout, or any of the setups which are listed under the Arrange options.

It is possible to set up to two Custom Layouts, for easy access, or to save to hard drive an infinite amount of setups. This can be convenient when there is more than one editor who uses the same system or if you find different layouts suitable for different aspects of working within the program.

To save a Window Layout you need to:

1. Set up the layout according to your needs on-screen.
2. Select the Window menu.
3. Scroll to Arrange.
4. Choose Save Window Layout.
5. To recall a Window Layout choose Restore Window Layout.
6. Navigate to the setup of your choice.
7. Click Choose.

Important Details about the Interface

It is worth having a good understanding of the interface of Final Cut Pro. This gives you the power to use Final Cut Pro to its full potential and to achieve a variety of editing tasks in many different ways.

Look to the extreme left of the Timeline – notice there are green radio buttons next to each of the tracks. These are monitoring buttons for video and audio. If you press the green button on any of the tracks you are effectively switching it off – this will deactivate the monitoring for that particular track and gives you the ability to mute the audio, or kill the video, at the flick of a switch.

At the base of the Timeline is a little speaker icon. Click this and you will see a selection of controls open in the Timeline next to the green

 monitoring buttons. These controls give you the means to quickly isolate an audio track for monitoring purposes. The speaker icon does the same as the green radio buttons, whereas by clicking the headphone indicator this will switch off all tracks except that which you have just selected. This can be more efficient than isolating each of the tracks individually.

 At the bottom left of the Timeline there is a symbol that looks like two mountains – this is called Clip Overlays. Later, as you get into the editing and sound mixing process, you will find this facility extremely useful for adjusting audio levels and setting the opacity of video clips.

 To the right of Clip Overlays are four little boxes. These boxes affect the size of the clips as they are displayed in the Timeline. This is useful for increasing the visual size of the clips if you are working with a monitor which is cramped for screen real estate.

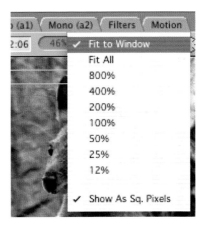

The following point is very important to take note of: look to the top of the Viewer and Canvas, just below the tabbed sections. You will notice there is a button with a percentage value in it. Click this button and it will reveal a series of numeric values – always keep this set to Fit To Window and have Show as Square Pixels checked at the bottom (assuming you are working with DV).

25

If you do not select Fit To Window you may encounter jerky playback and experience a great deal of frustration working out the solution. This applies to both the Viewer and the Canvas.

Considering that we have not even begun the editing process, the relevance of these details may seem a bit obscure at this stage. Rest assured it will make sense as you become familiar with the inner workings of Final Cut Pro.

Button Bars

Tools   Window
Audio Mixer
Frame Viewer
QuickView
Video Scopes
Voice Over

Button List

- ▶ File Menu
- ▶ Edit Menu
- ▶ View Menu
- ▶ Mark Menu
- ▶ Modify Menu
- ▶ Sequence Menu
- ▶ Effects Menu
- ▶ Tools Menu
- ▶ Window Menu

To speed up your workflow it is possible to add buttons to the top of the main windows of the interface. This enables you to quickly access functions which are often used.

1. Select the Tools menu and scroll to Button List. A list of assignable functions will appear.

2. Click any of the arrows to the left of each of the headings to reveal a list of the mappable functions.

3. Choose a function you wish to move to a button bar.

4. Drag the item from the Button List to the bar at the top of one of the windows of the interface. The button will then slot into place.

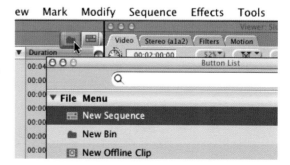

5. Press the button to perform the function assigned to it.

Customizing the Keyboard Layout

For editors migrating from other editing systems, this feature can make the learning curve with Final Cut Pro a much smoother transition. Every single button on the keyboard can be remapped to perform specific functions.

1	Select the Tools menu and scroll to Keyboard Layout.
2	Move right and choose Customize.
3	Click on the lock to unlock the keyboard.

Click any of the arrows next to the menu list to reveal a list of assignable functions.

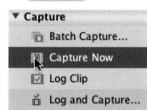

4	Choose a function and drag it to a key of your choice.
5	Press Save.

If you happen to change the settings and wish to revert back to the defaults then press Reset.

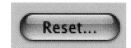

27

Note: Both settings, the button bars and the ability to customize the keyboard layout can also be accessed by Control clicking in the gray area next to each of the bars at the top of the windows.

Remove

Show Keyboard Layout
Show Button List

Save Main Button Bars...
Load Main Button Bars...

THE FOCAL EASY GUIDE TO FINAL CUT PRO 4

Why its meteoric rise? It worked right out of the box.
MICHAEL HORTON
LOS ANGELES FINAL CUT PRO USER GROUP

Nothing can be done with any editing program if you do not have the video material stored on the hard drives of your computer. It is therefore essential to capture the material before you can begin the editing process. Before you can begin capturing you must instruct the computer where the captured video files will be stored – in Final Cut Pro this area is called the Scratch Disks.

Setting Scratch Disks

Final Cut Pro provides the facility to set up to 12 scratch disks. While it is unlikely you will have 12 hard drives attached to your system the facility is there if you need it.

1 Go to the File menu which is found at the top left of the screen.

2 Scroll down to Log and Capture and release your mouse button. The Log and Capture window will now open.

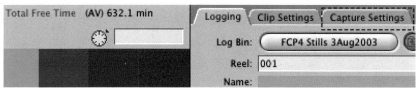

3 Click on the Capture Settings tab which is located to the right of the Log and Capture window.

4 Click the Scratch Disks button. This will reveal the Scratch Disks window.

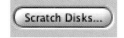

5 Press the Set button closest to the top – the reason there are several Set buttons is to allow one to set multiple scratch disks.

30

CAPTURE

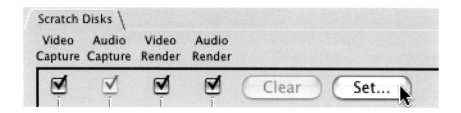

6   Navigate to the hard drives on your computer.

7   Double click a hard drive. This will set the hard drive as the first scratch disk in the list. If possible, select a drive that does not contain the operating system for your Mac (this ensures optimum performance when editing).

8   Alternatively, single click a hard drive of your choice and then press the New Folder button. Name the folder, for example Media Files, and press Create.

9   Press Choose and this will return you to the Scratch Disk window. Make sure you have Video Capture, Video Render and Audio Render

all checked with a cross. By default this will already take place for the top scratch disk.

You have now defined an area to store your video files.

If you double clicked a drive at point 7 three folders will have been created on that drive – one for video files, one for render files, and another for audio render files. If you moved on to point 8 you will have created a named folder inside of which the three folders exist.

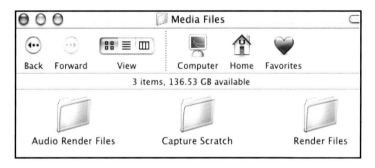

You can set up to 12 scratch disks depending on how many hard drives you have available. It is an advantage to set several scratch disks as Final Cut Pro automatically fills up the next available hard drive once the first hard drive becomes full.

Setting scratch disks will become second nature once you have run through the process a few times. The method described gives you the power to define exactly where you want your files to be stored. Providing you are disciplined you will be able to manage the media for each project you work on.

Working with Formats other than DV

Final Cut Pro has been designed to natively capture material which originates on either the DV or DVCam format. It can also capture footage which originates on DVC Pro 50.

When working with footage beyond the scope of DV and DVC Pro 50 you have two options. You can either transfer the original material to DV or DVCam for editing, or you can purchase a PCI card or external device designed to work with either standard or high definition footage.

The choice is up to you which option you wish to pursue. The option of working uncompressed or high definition may be quite alluring, however, be aware that the requirements to edit these formats will go beyond simply installing a card or external device to your Mac. There are certain hard drive requirements and configuration issues for working with formats outside the realm of DV. This can bump up the cost of setting up a system considerably.

These points considered, Final Cut Pro 4 provides a competitive and cost-effective solution for working with multiple standards of video or film. It is up to you whether you wish to move away from the simplicity and reliability of working exclusively with DV.

Methods of Capturing DV Footage

There are three ways to capture DV footage when using Final Cut Pro: **Capture Clip**, **Capture Now** and **Batch Capture**.

Capture Clip, as the name suggests, is used to capture a single clip at a time. It requires the editor to first mark 'in' and 'out' points. An 'in' point refers to the position on the tape where the capture process is to begin and the 'out' point is where the capture process is to stop. Once the 'in' and 'out' points are marked the computer cues up the tape in the deck/camera to the appropriate point and transfers the material onto hard drive.

Capture Now is used to capture clips 'on-the-fly'. This means the capture process begins the moment the editor instructs the computer to begin capturing and stops when the Escape button is pressed.

Batch Capture is used to capture multiple clips. Each clip is first 'logged' and the computer is then instructed to capture each of the clips in succession.

Deck Control

To capture video files to hard drive it is essential to know how to control the replay deck or camera from the computer. This is quite simple and has been well integrated into the editing interface. All operations are easily accessible using keyboard commands.

Space Bar	Play
J	Play Backwards
K	Stop
L	Play Forwards
i	Mark 'in' point
o	Mark 'out' point

Each of the play commands J and L work in increments. By pressing J or L up to five times will speed up the result. This will be obvious as we get further into the Final Cut Pro workflow.

The Capture Window

The Capture window is the facility provided within Final Cut Pro to enable you perform the capture process. It is important to understand the controls within this window and how to use them.

1. To open the Capture window first make sure your deck/camera is switched on. If you are using a camera make sure it is in VTR mode.

2. Choose the File menu at the top left of the screen. Scroll to Log and Capture.

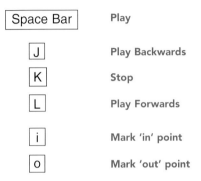

The Capture window will now open.

CAPTURE

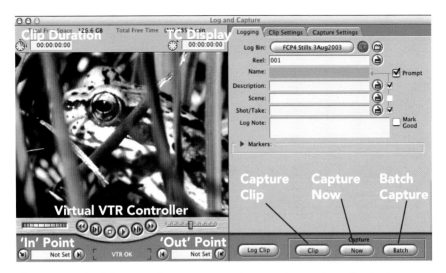

Above is marked the crucial areas one needs to understand to effectively use the Log and Capture window.

Clip Duration – when logging clips for Batch Capture or using Capture Clip, 'in' and 'out' points must first be marked. The duration of the clip is calculated by Final Cut Pro and displayed in the Clip Duration window.

Time Code Display – whenever a DV tape is playing, a running display will show the timecode numbers ticking over. If you stop the tape the timecode at the exact point where the tape is parked will be displayed.

'In' Point – an 'in' point is marked by pressing the letter 'i'. The marked 'in' point is displayed in this window.

'Out' Point – an 'out' point is marked by pressing the letter 'o'. The marked 'out' point is displayed in this window.

Virtual VTR Controller – just like most VTRs have stop, play and shuttle commands – this virtual controller performs similar functions.

Capture Clip/Capture Now/Batch Capture – used to perform the capture functions.

Also note the display at the top of the Capture window which tells you how much free space is available on your computer and how much this capacity

35

equals in minutes. The amount of space is the sum total available on the scratch disk or disks you set earlier.

You can therefore determine whether you have room on your hard drives to capture the material required for your project.

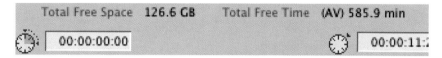

Capture Clip

If you wish to capture a single clip at a time this is easily achieved using the Capture Clip method. When using this method you need to first mark the 'in' and 'out' points for the clip you wish to capture.

1. Put a DV tape into your deck or camera and make sure it is switched on. If you are using a camera make sure it is in VTR mode.

2. Open the Capture window which is accessed through the File menu. Alternatively, press the Apple key and the number 8 and this will achieve the same result (the Apple key is located immediately left of the Space Bar).

3. Press the Space Bar and your deck or camera will spring to life. If it doesn't, press play on your deck/camera to engage the heads. From this point on remote control of your Firewire device will work direct from the keyboard.

Once the tape is playing at speed, the result of having pressed the Space Bar, you can then spool through the tape using the J K L method. As mentioned earlier, pressing the letter J will run the tape backwards, K is for stop (or use the Space Bar to start/stop the tape) and the letter L is to run the tape forwards. By pressing the letters J and L multiple times affects the replay speed incrementally. If you press the letter J once the tape will spool backwards at normal speed, press it again and the tape will continue backwards, however, slightly faster. Press it again and the speed will increase until the maximum speed is attained after five taps. Likewise, when using the letter L the tape will shuttle forward in increments until a maximum speed is achieved after five taps.

4   When you get to the point where you want the capture to begin press the letter 'i' – this will mark the 'in' point. Similarly, press the letter 'o' to mark the 'out' point. If you look to the bottom of the Capture window, the timecode reference for the marked 'in' and 'out' points will be displayed.

Note: when marking 'in' and 'out' points you can mark the points on-the-fly. This means you can mark 'in' and 'out' points by pressing the letters 'i' and 'o' at the appropriate points while the tape is running. If you prefer, while the tape is playing, hold down the letter 'i' and release it when you get to the point where you wish the 'in' point to be marked. The same applies for the 'out' point. Hold down the letter 'o' and release it to mark the 'out' point. Final Cut Pro is flexible in that the same result can be achieved in a variety of ways.

5   Once the 'in' and 'out' points have been marked press the Clip button at the bottom of the Capture Window. You will then be prompted to name the clip.

6   Name the clip and press OK. The Mac will then instruct the deck/camera to cue up the clip which will then be captured to disk and placed into the Browser for you to access.

By repeating this process you can capture as many clips as you wish.

Capture Now

An alternative way to capture clips is to use the Capture Now facility. This is a simple method that does not require you first to mark the 'in' and 'out' points.

1. Open the Capture window and play the tape in your deck/camera.

2. Press the Now button which sits immediately to the right of the Clip button in the Capture window. Immediately upon pressing 'Now' the capture process will begin. The images on the DV tape will be mirrored in a large window on your computer monitor. A message at the bottom of this window will confirm that capture is taking place.

> Capturing Clip – NOW CAPTURING (press 'esc' to stop)
> WARNING: Capture Now is limited to 30 min

3. Once the material you want has played, press the Escape key (located top left of your keyboard) to exit Capture Now. The capture process will stop and the clip will be placed into the Browser.

You can then name the clip by overtyping the name assigned to it by Final Cut Pro.

Always remember to close the Capture window once you have completed the capture process (do this by clicking the extreme top left of the Capture window). Failure to do so will prevent video and audio from playing through the Firewire – to your deck or camera – and onto your television monitor (assuming you are working with this configuration). By closing the Capture window this problem will be avoided.

Batch Capture

Batch Capture is an extremely useful facility for capturing many clips at a time. It is necessary to first log the clips you wish to capture, name them and then invoke the Batch Capture function.

1. Cue the tape in your deck/camera to the point where you wish to mark the first 'in' point. Mark the 'in' point and 'out' point for the first clip you wish to capture.

2. Press the Log Clip button – located at the bottom of the Log and Capture window.

3. You will now be prompted to give the clip a name. Do this and notice the clip appears with the name you assigned to it in the Browser window with a diagonal red line through it. The red line indicates that the clip is logged but not yet captured to disk.

4. Log as many clips as you wish to capture from the tape in your camera/deck.

5. Once you have logged the clips you wish to capture stop your DV tape by pressing the Space Bar or stop button on your deck/camera.

6. Look at the Browser area where your clips are logged – each clip will have a diagonal red line through it. Notice that the last clip you logged is highlighted. Click once anywhere in the Browser to deselect it. This is particularly important – otherwise when you try to Batch Capture, only the highlighted clip(s) will be captured.

7. In the Log and Capture window press the Batch button located bottom right. You will now be prompted with a screen full of information. Check that the Capture Preset corresponds with the format you are working with. If not, click the Capture Preset bar to select the format you are working with. Click OK to continue.

8. A window will now appear stating the number of clips which are ready to capture. Press the Continue button.

9 After a short pause the DV device will cue up the first of your clips and the Batch Capture process will commence. As each clip is captured the DV device will stop and shuttle to the next clip and so forth until all the clips are captured.

10 Once all the clips have been captured you will be prompted with a dialog box which tells you the numbers of clips that have been successfully captured. Click the finished button and, as if by magic, all the red lines will disappear from the clips in the Browser, signifying that they are stored on hard disk and accessible to work with.

Always remember to close the Capture window once you have completed the capture process or video and audio will not play through the Firewire cable to your deck or camera.

Selectively Capturing using Batch Capture

When you are working with Batch Capture you can be selective about what items you wish to capture. You can choose to capture Selected Items, Offline Items or All Items in the Logging Bin.

1 Highlight the items in the Browser which you wish to capture by holding down the Apple key and clicking on the individual clips.

2 Open the Log and Capture window.

3 Click the Capture bar at the top of the Capture window and choose Selected Items in Logging Bin.

4 Press OK followed by Continue.

Some users have reported problems when using the Batch Capture function. There are times when Final Cut Pro will attempt to capture everything in Browser, including material that is already captured to disk. This could possibly be due to the user not numbering reels correctly or trying to capture clips with matching timecode to previously captured clips.

Fortunately, there is a workaround which is 100% effective in these situations.

1. Log the clips you wish to capture so that you have a list of offline clips with red diagonal lines running through them.

2. Create a new project – select File – New Project.

3. Drag the tab of the new project so that it becomes a floating window.

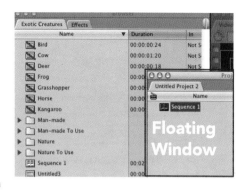

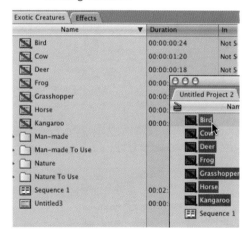

4. Drag the offline clips from your original project into the new project.

5. Invoke the Batch Capture command. The only clips that will be captured are those in the new project. The captured files will automatically relink to those offline clips in your original project once the capture process has finished.

This simple workaround will work flawlessly and give you guaranteed trouble-free capture. Thanks to Charles Roberts for saving me at 4am, one lonely London morning, when time was running out as a deadline rushed towards me!

Note: a similar method needs to be employed when working with OfflineRT. When you are ready to online your project you need to drag your edited Sequence into a new project and then use the Media Manager to create an offline at full online quality. Final Cut Pro will then only recapture the exact bits of media you need.

Getting the Most Out of the Capture Process

An important part of preparing for the editing process is to watch your material. To edit your footage properly you have to know what there is to work with. I always use the capture stage as a dual purpose procedure: (1) to watch the footage, properly, in real time as it is captured and (2) to capture the footage onto hard drive.

In the film world, as soon as the footage made it back from the lab, the dailies would then be screened. A projector was laced up and the illuminated images thrown forth in a display of light onto the screen. Based on what the director saw the rest of the shooting could then be arranged.

You are the director, the Log and Capture facility is your projector. So watch your images and build the structure of your film in your mind before you cut a single shot to another shot. You decide what to take in, what to leave out, and what to reshoot if necessary.

Importing Music from CD

Films are primarily made up of sound and picture. Much of the sound is recorded at the same time the picture is recorded. When it comes to music, more often than not, this will be sourced from compact disc.

Importing tracks from CD into Final Cut Pro is straightforward. To begin with you need to hide the Final Cut Pro application (press Apple H) and go to the desktop to access the CD.

1. Insert your CD into the Mac's CD/DVD drive.
2. Double click the CD icon on your desktop to open it.
3. Drag the track/s you wish to work with direct from the CD to the desktop – wait while the copy process takes place. You may wish to rename the CD track/s once the copy process is finished.

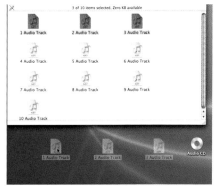

Once the track or tracks have been copied, you need to go back into Final Cut Pro to import the tracks.

1 Make sure Final Cut Pro is open in front of you.

2 Select the File menu and scroll down to Import. Scroll right and select Files.

3 Navigate to the desktop and locate the track/s you wish to import.

4 Highlight the track or tracks you wish to import. If you want to import more than one track hold down the Apple key and click each of the tracks with your mouse button.

5 Press the blue glowing Choose button.

The CD track/s will now appear in the Browser and these are represented by a speaker icon. Rename the track/s if you choose.

An alternative way to import files into Final Cut Pro is to drag them direct from the desktop into the Browser. This will achieve exactly the same result as using the Import Files command.

It is important now to convert the audio sample rate of the imported track/s to match the rest of the audio in your project.

Converting Audio Sample Rates

It is easy to import CD tracks into Final Cut Pro – the complicated part of the process is to get the CD sample rate to match that of the rest of your project. This is important. Mismatching sample rates can cause a variety of problems including pops, clicks and sync drift.

You will remember DV audio is recorded at either 16 bit – 48 kHz or 12 bit – 32 kHz. The key to trouble-free audio editing within Final Cut Pro is to make sure that all audio is of the same sample rate. Commercial CDs are recorded at 44.1 kHz. It is therefore advisable to convert the sample rate of the CD track to match the rest of the audio in your project.

1 Highlight the CD audio track which needs to be converted in the Browser.

2 Select the File menu, scroll to Export and select Using QuickTime Conversion.

3 Click on the Format bar to reveal a list of options.

4 Select AIFF.

5 Click the Options button to the right.

This will reveal a series of settings. Where it reads Rate – click the two arrows facing in opposite directions. This will reveal a list of audio sample rates.

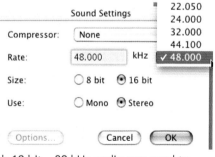

6 Set the sample rate to 48 kHz and check that 16 bit and Stereo are selected. Click OK. (If you are working with 12 bit – 32 kHz audio you need to select the 32 kHz setting.)

7 Name the file and save it to hard disk. The conversion process will now take place.

8 Go to the File menu and select Import. Locate the CD file you just exported and import this into the Browser. The sample rate has now been converted to 48 kHz and will now match the rest of your project.

*Final Cut Pro's benefit is that it is more productive.
It allows the editor to get more done.*
MICHAEL VITTI
NEW YORK FINAL CUT PRO USER GROUP

Viewing Clips

Now that you have captured your clips you need to be able to view them. This is the first step towards sorting through your footage. To view your material double click any of the clips in the Browser and immediately the clip will open in the Viewer. Press the Space Bar and the clip will play.

The controls in the Viewer window can also be used to play the clip. You can move quickly through the clip using the yellow Scrubber Bar, located below the image in the Viewer. Simply click once with the mouse and move the Scrubber Bar backwards or forwards.

It is also possible to shuttle through the clip using the J K L method. By tapping the 'J' or 'L' key up to five times the speed will increase in increments. Press 'K' or the Space Bar to stop. Press the Space Bar again to play.

To jog through the clip a frame at a time press the horizontal arrow keys located to the right of the Space Bar. The left arrow takes you backwards a frame at a time while the right arrow takes you forward a frame at a time. Hold down the Shift key and press either of the arrow keys and this will move forwards or backwards through the clip a second at a time.

Playing Video through Firewire

It is desirable to play the video signal through a deck or camera and onto a standard television screen. This is because the images on a television screen provide a true representation of the final quality of your finished movie. Otherwise, you will be working exclusively off the computer monitor which provides a different type of picture to that of a television set. The fact that the final product will most likely be viewed on a television means that it is desirable to view the material on a TV while you work.

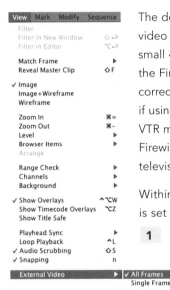

The deck or camera must be set to receive video through the Firewire cable. A large 6-pin to small 4-pin Firewire cable will link your computer to the Firewire device. You must make sure the correct input is selected if you are using a deck, or, if using a camera, make sure that it is switched to VTR mode. Beyond this the output from the Firewire device needs to be fed into the television set.

Within Final Cut Pro check that the External Video is set to play all frames through Firewire.

1   Choose the View menu found at the top center of the screen.

2   Scroll to video and select All Frames.

The video signal will now play direct from the Mac to your deck or camera and onto the television set.

DV Start/Stop Detection

Back in the old days the film editor would take the workprint when it returned from the lab and cut it into pieces. These pieces were individual shots or sequences of film. The problem with having a huge amount of film on a reel, which had not been broken down into shots, was that it was unwieldy and time consuming to work with. Imagine if the editor of 40 years ago had a machine that would do that part of the process for them.

Well you do have it. Inside of Final Cut Pro is a remarkable feature that will break up long captured clips into individual shots. Each time the record button is pressed on a DV camera a signal is recorded to the tape – another signal is then recorded to tape when the recording stops. Final Cut Pro recognizes this recorded signal and once footage has been captured you can run your footage through what is called Scene Detection. Effectively Scene Detection scans through your footage and breaks it up into individual shots.

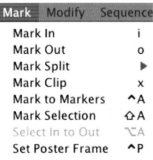

1   Highlight a clip or group of clips in the Browser (a group of clips can be selected by dragging a lasso with your mouse around several clips or by holding down the Apple key while clicking on the individual clips).

2   Select the Mark menu and scroll down to DV Start/Stop Detection – release your mouse button.

3   A progress bar will appear as the clip or clips are scanned by the computer.

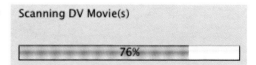

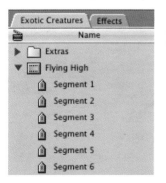

4   Look to the clip(s) you have just scanned. Notice there is a little arrow next to the clip name. Click the arrow and this will reveal a set of pink arrows. Each of these arrows represents a time when the DV camera has started and stopped recording. If you click any of the segments you can then view each of the shots in the Viewer.

ORGANIZING YOUR FOOTAGE

DV Start/Stop Detection is a very useful function. If you wish you can capture an entire tape's worth of material and then have the computer break up the footage into individual shots. Effectively the computer does a great deal of sorting through your raw material for you. What it doesn't do is name the individual shots or sort out the good takes from the bad. That is something you must do.

Working with Bins

In the old days before videotape was invented, and certainly before digital cameras and computers were used to acquire and edit productions, a film editor would organize strips of film in an area known as a trim bin. These film strips were hung on a horizontal rack and ordered according to the wishes of the film editor.

While a lot has changed technologically, when working with a non-linear editing system such as Final Cut Pro, it is still crucial to order your material. Otherwise, it soon becomes impossible to track down your shots, particularly if you are working with hours of footage and thousands of clips. Final Cut Pro certainly has the power to handle productions of this magnitude!

To facilitate a simple way of ordering your material it is possible to create what are called bins within the Browser window. Within each of these bins you can store individual clips. The term bin, as you may have guessed, is taken from the era of film editing.

1. To create a new bin select the File menu at the top left of the screen – scroll down to New and select Bin. Alternatively, press Apple B (the Apple button is located immediately left of the Space Bar).

2. In the Browser a box will appear titled Bin 1. This box is clear and different in shape to the clips so there is little likelihood of confusion. You can rename the bin by typing a name immediately after it has been created. Should you wish to rename the bin later, simply click

49

once on the text area, then pause, and click again in the text area. You can now overtype the title and name the bin whatever you wish. Press return once you have renamed the bin.

Now that you have created and named a bin you can place clips inside it. Clips can be moved, one at a time, by clicking once on the clip and dragging into a bin with the mouse. To select multiple clips use the mouse to drag a lasso around the clips you wish to highlight. Drag the highlighted clips over a bin and release the mouse button. The clips will then be dropped inside the bin.

Highlight Clips and Drag these into a Bin

Several clips can also be selected, one at a time, by holding down the Alt/Option key (located second to the left of the Space Bar) and clicking on each of the clips you wish to highlight. Drag the highlighted clips into the bin and release your mouse button.

To view the contents of a bin:

1. Click the triangle to the left of the bin's title and the contents will be displayed in descending order.

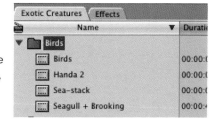

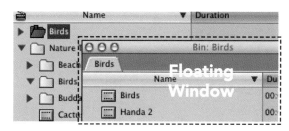

2. Alternatively, double click the bin and a floating window will appear with the bin's contents visible in front of you. To close the floating window click the button at the top left and the bin will return to its original position within the Browser.

ORGANIZING YOUR FOOTAGE

You can create as many bins as you like. And you can also store bins within bins. Simply drag a bin over another bin and release your mouse button. The result is a bin stored within a bin.

Items can be moved from one bin to another by highlighting and dragging. Should you wish to move an item from within a bin back to the Browser, you must double click the bin to open it as a floating window and then drag the item or items out of the bin and into the Browser where they will be positioned.

If you want to delete either a clip or a bin highlight the item and press the delete key. Note that the items are only deleted from the Browser and not from your hard drive. Everything inside of Final Cut Pro works by referencing to the original files which exist in the scratch disk folder/s which you set up earlier. Original clips remain stored on the hard disk of your Mac unless you actually go into the hard drive, remove the items and then place them in the trash on the dock. By emptying the trash the items are then deleted.

Working in Icon View

By default Final Cut Pro will display clips in list mode, which means the clips are represented by the names you give to them during the logging process. It can be advantageous to view your clips as icons, or miniature pictures, particularly if you are the sort of person who prefers to work visually. It can be easier to identify a clip by a picture icon, rather than scrolling through an alphabetical list of words.

To view clips in icon mode:

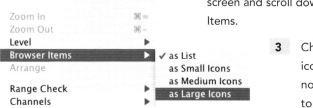

1   Click in the Browser to make the Browser active.

2   Select the View menu at the top of the screen and scroll down to Browser Items.

3   Choose any of the icon views – you will notice it is possible to view as Small,

51

Medium or Large Icons. The items in the Browser will now be represented by pictures, rather than by words.

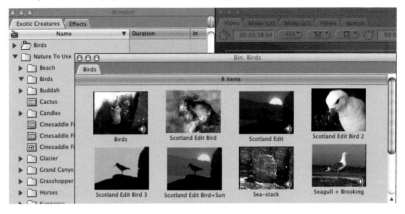

By selecting the Arrange function under the View menu this will line up the icons alphabetically.

My preference is to keep the items in the Browser in list view and items inside each of the bins in Icon view. By using this combination one has the advantage of being able to view the material both ways. If one clicks on the arrow to the left of any of the bins, the clips will be displayed in list view; whereas if one double clicks a bin the contents can then be displayed as icons.

Setting Poster Frames

As described, when you work in icon view, each clip is represented visually in the form of a miniature icon. The picture used to represent each of the clips is called the Poster Frame. The Poster Frame is determined by the first frame of the clip.

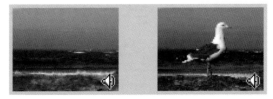

A problem arises when the first frame of the captured clip is not representative of the shot. Look at the above example. The image on the left does nothing to represent a Seagull, whereas the image on the right says it all.

It is possible to set any image from within the clip to be the Poster Frame.

1 Open the clip into the Viewer and position the Scrubber Bar on the frame you wish to display.

2 Select the Mark menu at the top of the screen and scroll down to Set Poster Frame. Release the mouse button and the image on the thumbnail will now change to that which you have selected.

It is also possible to reset the Poster Frame to the first frame of the shot simply by selecting the Clear Poster Frame command which is also found under the Mark menu.

Searching for Clips

Final Cut Pro is a powerful editor capable of referencing to thousands of clips stored on the hard drives of your computer. As the editor you have to know what footage is there and how to get to it. It is all very well to know that it is there somewhere – if you can't find it you are lost.

The most useful database of all is the human mind. An editor will constantly refer to the list in their mind to retrieve a shot ephemerally before actually doing so electronically.

When working on a large project, with hundreds or thousands of separate clips, you need a system to find what you need. Providing you have taken care to label each of your clips in a way that is easy for you to identify with, you will then be able to search for any clip in your project. Apple has made this possible in an extremely simple and elegant way.

To search for a clip:

1 Click once anywhere in the gray area of the Browser.

2 Hold down the Apple key and press the letter F. This will open the Search/Find dialog box. Alternatively, this can be accessed from the Edit menu by scrolling to the Find command.

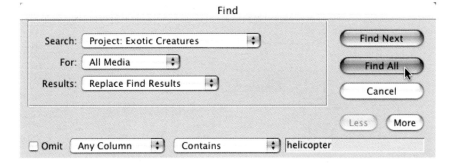

3. Type the name of a clip or part of the name you wish to search for and then press Find All. All relevant items will then be placed in a bin in front of you.

What would have taken a lot of effort in the past, when working with film or tape, has now been reduced to only a few seconds' work. You still need to rely on your mind to know what you are looking for, however, the difficult work of actually locating the clips has been made very easy. That is assuming you have labelled your clips correctly and that you know what you are looking for in the first place.

The Search facility within Final Cut Pro can be quite sophisticated. There are many options available beyond the simple method of searching within the Browser as just described. It is possible to search by name, column, through the log notes and a variety of other criteria.

> *What's so good about it is it is uncompromisingly powerful. The most insanely powerful application for under a thousand bucks.*
> STEVE MARTIN
> APPLE CERTIFIED TRAINER

Any film is literally built. Just as a novel will have chapters and subplots, each and every film has an underlying structure. The raw components needed to build the film are planned for in the scripting stage, gathered while filming and structured during editing.

I liken the filmmaking process to making a set of chopsticks from a tree-trunk. An entire tree can be whittled away to leave nothing remaining other than two small pieces of wood – these are the chopsticks. Film or video is the same. A mountain of footage is acquired and throughout the editing process this footage is chopped down to a fraction of its original size to leave a small, yet refined, remnant of the original content.

In simple terms editing is nothing more than putting shots and sounds together. In reality it is much more than this. It is both a technical and a creative process – it is also intuitive. Anyone can string words together, but not everyone is able to write a good story or a good book without a sound knowledge of language. Editing is similar.

There are several key methods of editing with Final Cut Pro. Of these the most important to understand are Insert and Overwrite Editing. While there are other methods available such as Replace Editing, Fit to Fill and Superimpose, provided you understand Insert and Overwrite Editing you will be able to edit any production.

Let me stress, the key to editing with Final Cut Pro lies in the difference between Insert and Overwrite Editing and when to use one or the other. Beyond this, you must understand how to control audio in relation to these two ways of editing. Once you have this clear in your mind you are well on the way to mastering Final Cut Pro. You will then have the technical knowledge to make a film that is fully professional and equal to whatever you watch on TV or see at the cinema. I'm not kidding here. You will be able to edit anything from a thirty second commercial to a feature film.

Insert and Overwrite Editing

Think of the old days when film was edited in a cutting room. The editor would take two pieces of film, line them up in a splicer and join them together. As more pieces of film were cut together a Sequence was formed. As more Sequences were crafted these were joined together to build completed scenes until finally titles and effects were added. Once all the scenes were completed the final result was a finished film.

When putting the pieces of film together the editor had two choices: either a piece of film was added to the shots already cut together and therefore the overall length of the Sequence was increased, or a piece of film was placed into the Sequence and a corresponding amount of film, the same in length, removed – thus the overall duration did not change. These two choices are what Insert and Overwrite Editing are all about.

When you build your movie in Final Cut Pro you edit various shots together. Whenever these shots are put together you must decide whether you are adding a shot to a Sequence and therefore increasing the overall length of the movie, or, whether you wish to simply replace a section with another shot previously not included (thus keeping the Sequence the same in length).

When editing with a non-linear system such as Final Cut Pro the editor has a lot more in common with the film editors of yesterday than the tape editors of recent times.

Getting Started with Editing

1. Check that you have a Sequence open. If you can see the Timeline in front of you then a Sequence is already open. Your Sequences are stored in the Browser, the same area where your clips and bins are kept. If a Sequence is not open double click a Sequence in the Browser and the Timeline will appear.

2   Choose a clip from the Browser and double click it – this will load the clip into the Viewer (you may have to open one of your bins if you have filed away all of your clips).

3   With your clip loaded in the Viewer scrub through it. You can do this either by using the Scrubber Bar, or by using 'J' to scrub backwards or 'L' to scrub forwards (tap either of these keys in increments to speed up the rate of scrubbing). Use the Space Bar to start or stop.

4   Choose a point in the clip where you wish to mark an edit point. Press the 'i' key to mark the 'in' point.

5   Choose the point where you wish to mark the end point of the clip – this will be the 'out' point. Press 'o' to mark the 'out' point.

The procedure for marking 'in' and 'out' points is the same as that you already experienced during the logging process.

Once you have marked the 'in' and 'out' points you are now ready to edit the shot into the Sequence. There should be no shots in the Sequence at this stage and therefore the Timeline will be empty.

Make sure the Scrubber Bar is positioned at the beginning of the Timeline. To do this click anywhere near the numbers in the light shade of gray at the top of the Timeline. You will now see the yellow Scrubber Bar with a vertical line extending from top to bottom. Drag the Scrubber Bar along this light gray area and position it at the beginning of the Sequence (all the way to the left). Alternatively, press the Home key on your keyboard and this will have the same effect.

6 Click with your mouse in the center of the clip that you have loaded into the Viewer. A small transparent box will appear where you click. While still holding the mouse button, drag the cursor over the Canvas (the window to the right of the Viewer).

A selection of options will appear. The top option is Insert, followed by Overwrite, then Replace, Fit to Fill and Superimpose. At this stage we are only concerned with the first two options: Insert and Overwrite.

7 Move the cursor, with the transparent box, over the Insert button (marked yellow). Release your mouse button. Look to the Timeline and notice there is now a single block positioned at the beginning. This is the first shot of your Sequence.

8 Repeat the above process with another shot. Double click a shot to load it into the Viewer and mark the 'in' and 'out' points. Click on the shot in the Viewer, drag this over the Canvas and release it over the yellow Insert button. You now have two shots in the Timeline.

9 Edit several more shots together – choose between five and ten shots. When you have cut these together, use the Scrubber Bar in the Timeline to move back and forth through the Sequence. Position the Scrubber Bar at the beginning of the Sequence and press the Space Bar. The shots will play in the Canvas and onto your television monitor – assuming you are plugged into a deck/camera and video is playing through the Firewire. If video is not playing through the Firewire check that All Frames is selected under External Video found under the View menu.

Note: you can also scrub through your Sequence in the Canvas by using the Scrubber Bar at the bottom of the Canvas. The Canvas and the Timeline are linked in that the Timeline is a graphical representation of all the shots edited together in the Canvas. The Timeline shows individual clips as blocks, whereas the Canvas shows the shots as moving images.

Distinguishing between Insert/Overwrite

In the Timeline you should now have several shots edited together. Position the Scrubber Bar in the Timeline at the beginning of the Sequence. Press the upward arrow on your keyboard (located to the right of the Space Bar) and you will find you are now able to skip forward between each of the shots. Press the downward arrow and you will find you can skip backwards through your shots, one by one.

 Skip Forwards   Skip Backwards

Now, position the Scrubber Bar in the middle of the Sequence.

| 1 | Open a shot in the Viewer and mark the 'in' and 'out' points. |

2   Drag this shot over to the Canvas, however, this time, instead of releasing it over the Insert button, position it over the Overwrite button (marked red). Now release your mouse button.

3   The shot will be edited into the Timeline – but it will not push all of the other shots in front of it further along in the Sequence. Instead, it will write over a portion of the Sequence beginning where your Scrubber Bar is positioned.

If it is not obvious that this has happened it may be necessary to condense the overall spread of the shots on the Timeline. To do this, look to the bottom of the Timeline and find the slider bar with two ribbed ends. Drag either of these ribbed ends and you will see that the Timeline can be expanded or contracted. This does not affect the length of your movie in any way. What it does is to affect the display of your Sequence.

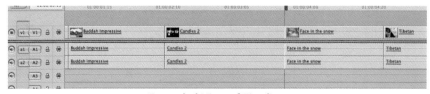

Expanded View of Timeline

EDITING

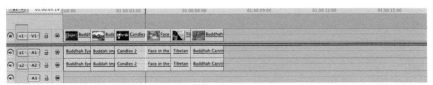

Contracted View of Timeline

This is very useful when you have a Sequence that is long and you wish to be able to view the entire Sequence on the screen in front of you. It is also useful when you wish to expand the Sequence for fine control to allow precise positioning of the Scrubber Bar.

To make completely clear the difference between Insert and Overwrite Editing it is advisable to condense the Timeline so the entire contents are visible on screen. You will then be able to determine the type of edit: if the Timeline has been made longer, you have performed an Insert Edit; if the length does not change you have performed an Overwrite Edit.

To be able to see the difference between Insert and Overwrite Editing:

1. Position the Scrubber Bar on a shot change in the middle of the Timeline.

    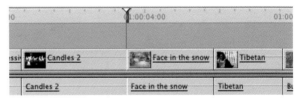

2. Open a shot in the Viewer and mark an 'in' and 'out' point.

3. Drag the shot from the Viewer to the Canvas and position it over the Insert button. Observe the Timeline as you release the mouse button and notice all other shots get pushed further along the Sequence.

4. Hold down the Apple key and press the letter 'Z'. This will undo the action you have just performed.

63

5 Repeat the procedure of dragging the shot from the Viewer to the Canvas, however, this time, release it over the Overwrite button. It should be apparent that a different effect has taken place. The shots in the Sequence are not pushed further along the Timeline – they all stay in exactly the same position. What has happened is that the shot you have just edited into the Sequence has written over a portion of the Timeline. The length is determined by the 'in' and 'out' points in the Viewer.

If you look at the top left of the Viewer you can see the duration of the shot you are working with. This is measured in seconds and frames. If you change the position of either the 'in' or 'out' points Final Cut Pro will calculate the new duration.

Note: it is not necessary to drag the video from the Viewer to the Canvas to perform an Insert or Overwrite Edit. If you prefer, mark the 'in' and 'out' points in the Viewer and press the yellow or red button at the bottom of the Canvas. Providing you remember that yellow is for Insert and red is for Overwrite then these functions can be accessed in this way.

Three Point Editing

So far we have only marked 'in' and 'out' points in the Viewer with the positioning of the Scrubber Bar determining where the Insert or Overwrite Edit will be edited in the Timeline. It is also possible to enter the 'in' and 'out' points directly into the Timeline. Simply position the Scrubber Bar where you want to mark the 'in' point and press 'i' and similarly press 'o' where you want to mark the 'out' point.

By marking a single 'in' point in the Viewer you can then perform an Insert or Overwrite Edit. The positioning and duration of the edit is determined by the 'in' and 'out' points marked in the Timeline.

EDITING

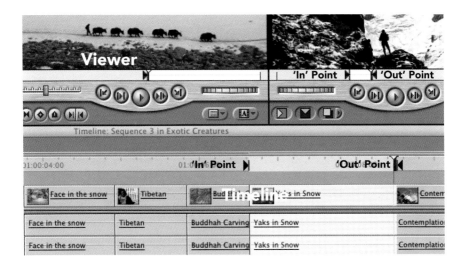

It is also possible to mark the 'in' and 'out' points in the Canvas.

What is being illustrated here is known as **Three Point Editing**. Essentially, all editing in Final Cut Pro works according to the Three Point Editing system. Even if it appears that only two points have been marked, the positioning of the Scrubber Bar in the Timeline serves as the third point.

It is important to be aware that whatever points are marked in the Timeline will be reflected in the Canvas and vice versa. The Timeline and Canvas are intimately related at all times – they are in no way independent of each other.

It may be clear at this stage just how closely the Viewer and Canvas mimic a traditional two-machine editing suite. If one forgets about the Timeline for the moment, all that is taking place is marking 'in' and 'out' points in the Viewer and/or Canvas. This is the same process as marking 'in' and 'out' points in a two-machine edit suite with a source VTR and a record VTR.

Other Editing Options

So far we have looked at Insert and Overwrite Editing. You will have noticed other options can be chosen when one drags a clip from the Viewer to the Canvas.

65

Replace Editing – this is used to overwrite a shot into the Timeline, from the Viewer, with the duration being determined by the shot which already exists in the Timeline. By marking an 'in' point in both the Viewer and Timeline/Canvas, the shot being edited will match the duration of the shot being replaced in the Timeline. There is no need to mark an 'out' point.

Fit to Fill – Four points need to be marked to achieve a Fit to Fill edit: an 'in' and 'out' point in the Viewer and an 'in' and 'out' point in the Timeline or Canvas. The shot in the Viewer will then be either sped up or slowed down to fit into the space of the shot which is being overwritten in the Timeline. The overwritten section of the Timeline will then need to be rendered (dealt with later).

Superimpose – This is used for creating a second layer of video. When using this type of edit a shot is edited from the Viewer to the second video track in the Timeline. No apparent result will be noticed, other than that of an overwrite edit, unless the edited shot is reduced in size, effectively creating a picture in picture.

Modifying 'In' and 'Out' Points

If you wish to clear 'in' or 'out' points there are several ways to achieve this.

1   Select the Mark menu at the top of the screen.

2   Scroll down and choose the relevant option: clear 'in' and 'out', clear 'in' or clear 'out'.

Keyboard shortcuts can be used to perform these functions. Hold down the Alt/Option button (two keys to the left of the Space Bar).

Alt/Option + x	**Clear 'in' and 'out'**
Alt/Option + i	**Clear 'in'**
Alt/Option + o	**Clear 'out'**

By holding down the Control key and clicking in the area where one scrolls with the Scrubber Bar, in either the

EDITING

Viewer, Canvas or Timeline, a contextual menu will appear. 'In' and 'out' points can be set or cleared by choosing the relevant command.

It is also possible to alter the 'in' or 'out' points by dragging or repositioning.

1 Click on the 'in' or 'out' point symbol in the Viewer, Canvas or Timeline and drag it to where you want it to be repositioned.

2 Alternatively, position the Scrubber Bar where you want the 'in' or 'out' point to be and press 'i' or 'o'. The 'in' or 'out' point is then repositioned and the previous 'in' or 'out' point is effectively deleted.

Directing the Flow of Audio/Video

The editing that we have done so far has involved editing video and audio at the same time. To produce a professional film one needs to be able to edit video and audio separately. This is easy to achieve with Final Cut Pro and as with many of the editing functions there is more than one way to go about it.

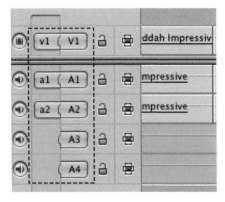

Look to the left of the Timeline and you will notice that where each track is labelled V1 (video 1) and A1 and A2 (audio 1 and 2) there is also a lower-case v1, a1 and a2 symbol. I refer to these as break-off tabs. If you click these tabs you will notice that they break away from the fixed video and audio symbols.

This is a simple patch facility which enables you to quickly and easily direct the flow of audio and video.

1 Click the break-off tab, v1, next to the capitalized V1 symbol.

67

2 You will notice it immediately slides slightly to the left, and is effectively broken away from the fixed V1 symbol. This means that track is inactive for editing.

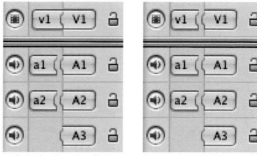

Click the Break-off Tabs to Target or Disable Tracks for Editing

3 Do the same to the break-off tabs next to A1 and A2. The lower-case tabs will slide to the left of the capitalized A1 and A2 symbols.

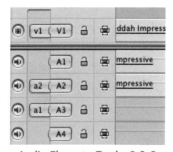

Audio Flows to Tracks 2 & 3

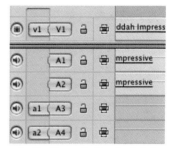

Audio Flows to Tracks 3 & 4

By clicking again on any of these tabs they will slide to the right and rejoin with the symbols to the right, indicating that those tracks are now targeted, or active, for editing.

Now, click one of the audio break-off tabs a1 and slide it to track A3. Release your mouse button and you will see that it remains joined to the new track where you have repositioned it. Repeat the procedure again with the audio break-off tab a2 so that it attaches to A4.

Wherever the break-off tab is positioned indicates that audio or video will flow to that particular track. If a tab is disconnected, nothing will flow to the track. The tabs toggle back and forth as they are clicked, indicating that a particular track is targeted or not targeted for editing. The tabs can be moved from one track to another by sliding, or by clicking, to allow you to direct the flow of audio or video.

For example, if you have V1 connected and A1 and A2 disconnected, then video will be edited across into the Timeline and audio will be restricted. In contrast, should A1 and A2 be connected and V1 disconnected – then only audio will be edited and the video will remain unaffected.

In simple terms, whatever is connected will be edited into the Timeline and whatever is disconnected will remain unaffected.

Locking Tracks

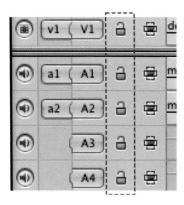

An alternative way to edit video and audio independently of each other is by locking or unlocking tracks. This provides a very simple and effective way to prevent either audio and video or a combination of both from flowing through to a particular track or series of tracks. It is as simple as locking the track or tracks that you do not wish to alter.

Look to the left-hand side of the Timeline and you will see there is a single video track and four audio tracks. This is the default number of video and audio tracks which Final Cut Pro provides you with when you launch the program.

To the immediate right of the V1, A1 and A2 symbols are little locks. Click on the locks and notice the track or tracks become grayed out.

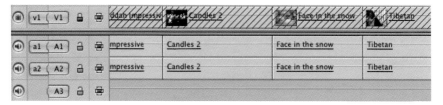

This means that when you edit material from the Viewer to the Canvas, and into the Timeline, the only part of the Timeline which is affected is that which is not locked (or not grayed out).

To lock a track prevents it from being affected during editing. The only way to reactivate the track is to unlock it. This is done by clicking the lock on the left-hand side of the Timeline. Once the track is unlocked it no longer appears grayed out. The usual rules regarding Insert and Overwrite Edits apply.

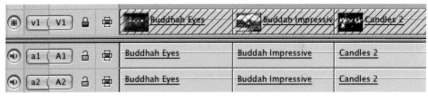
Video 1 Locked Audio 1 & 2 Unlocked

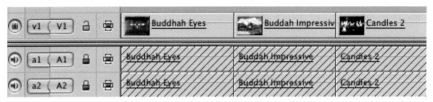
Video 1 Unlocked Audio 1 & 2 Locked

Adding and Deleting Tracks

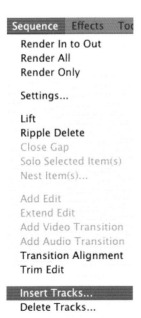

Final Cut Pro allows you to work with up to 99 layers of video and audio tracks. The default setup is a single video track and four audio tracks.

There are two ways to add or delete video and audio tracks to your project.

1. Select the Sequence menu and scroll down to either Insert Tracks or Delete Tracks.

2. A menu will appear giving

you the option to choose the number and type of tracks you wish to add or delete. You need to specify where these tracks are to appear in the Timeline by clicking the button options.

An alternative and easier method is to Control click in the gray area to the left of any of the tracks which already exist. A menu will open giving you the choice to either add or delete a track. You can Control click above, below or to the left of the tracks in the Timeline to open up the contextual menu which allows you to add or delete tracks.

Essential Editing Tools

A very important part of the interface that we have not dealt with so far is the Toolbar. With the layout I use this is positioned to the left of the Timeline. However, it really doesn't matter where it sits on the screen providing you can readily access it.

There are nine tools available, however, generally, I use only five of these for most editing tasks. There are also other tools hidden in the submenus within the Toolbar giving a total of 22 options in all.

Pointer (Select Item)
Edit Selection
Arrow (Select Track)
Roll Edit
Slip Edit
Razorblade
Magnifier
Crop
Pen Tool

Pointer (Select Item) – I call this the home tool – this is the tool I always have selected during the editing process. The Pointer is used for selecting and moving clips around in the Timeline. If I need to access the functions of the other tools I will choose another tool, use it, and then click on the Pointer again. By always having the Pointer selected you know where you are at all times.

Arrow (Select Track) – this is used for selecting individual or multiple tracks, or the entire contents of the Timeline.

Razorblade – used for cutting clips into pieces. Great for trimming edits.

 Magnifying Glass – most useful for expanding and reducing the Timeline. Useful for homing in on the exact part of a clip you wish to work with.

 Pen Tool – essential for adjusting audio levels. Also used for adding keyframes, thus allowing you to plot points over time. Useful for creating effects and adjusting video levels.

It is essential to understand how these tools work in order to edit efficiently. While shots can be strung together without ever touching the tools, in order to be able to trim edits, move shots around, home in on an exact part of a clip with absolute accuracy and to adjust clip levels and mix audio, one must be able to grasp these tools and how they can be used in combination with each other.

Undo/Redo

As your skills develop and you experiment with the facilities in front of you there will be times when you will get ahead of yourself and you will need to backtrack a few stages.

This is easy to achieve in the form of undo. It is also useful to be able to redo any of the actions you have performed.

At any time an edit can be undone by holding down the Apple key and pressing 'Z'. To perform a redo command press Apple + Shift + Z.

> **Apple + Z – Undo Action**   **Apple + Shift Z – Redo Action**

Multiple levels of undo can be achieved by pressing **Apple 'Z'** several times. Similarly, press **Apple Shift 'Z'** as many times as you wish to perform multiple redos. This ability to undo and redo is particularly useful when comparing changes in different edits. The number of levels of undo/redo can be set in User Preferences found under the Final Cut Pro menu located top left of screen. The default amount is 10 levels of undo/redo. This can be set to a maximum of 99.

Linked/Unlinked Selection

By default your audio and video are locked together. This means clips positioned in the Timeline will be married together in a similar way to film images and magnetic sound-striped tape running together in a synchronizer or a projector.

To illustrate the meaning of Linked Selection make sure you have selected the Pointer from the Toolbar. Check that you have several clips in the Timeline.

1. Point your cursor at a clip in the Timeline and click once. The clip is now highlighted.

2. While still holding down your mouse button, slide this clip to the right or left. Notice that both the audio and video move together.

3. Release the clip you are moving over one of the clips in the Timeline and the video and audio will overwrite the clip over which it is positioned.

4. Press Apple 'Z' and the Overwrite Edit will be undone. The clip will return to its previous position.

5. Click once on the green symbol which resembles a diagonal figure 8 inside a box, located top right of the Timeline. The symbol will turn gray and white to represent that Linked Selection is turned off.

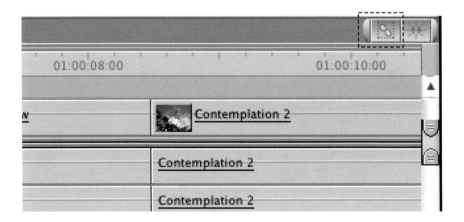

Linked Selection can also be accessed from the Edit menu at the top of the screen.

Scroll down to Linked Selection; if there is a tick, this means Linked Selection is on; no tick and it is switched off. Linked and Unlinked Selection toggles on and off when accessed through the menu.

With Linked Selection switched off repeat the procedure of selecting a video clip. Slide the video to the right and observe that while the video moves, the audio stays where it is. Conversely, select the audio and you can move this without affecting the video.

Video Moved
Independent of Video

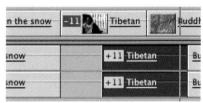

Audio Moved
Independent of Video

To select more than one track at a time hold down the Shift key, while using the Pointer tool, and items can be grouped together. Release the Shift key and the grouped items can be moved and repositioned wherever you wish.

To switch Linked Selection back on, click once on the white circles on the top right-hand side of the Timeline. These will turn green indicating that Linked Selection is switched on – this will apply even if you have moved video and/or audio independently of each other.

Note: an identical effect to linking or unlinking can be achieved by locking your tracks. Simply lock the tracks you do not wish to alter and then slide the video or audio of the clip you wish to move. Even though Linked Selection may still be turned on a locked track or tracks will override the link.

Moving Edits in the Timeline

You may have noticed when you use the Pointer tool to slide a clip to a different location the effect is that of an Overwrite Edit. It is also possible to move edits around in the Timeline, using the Pointer tool, and at the same time perform an Insert Edit.

To perform an Insert Edit within the Timeline it is crucial to press the keys in the correct order.

1 Using the Pointer tool highlight the clip you wish to move and release your mouse button.

2 Press and hold down the Alt/Option key and click once again with the Pointer tool on the clip you wish to move. Reposition the clip by dragging and release your mouse button at the point where you want the clip to be inserted in the Timeline.

This time the result is that of an Insert Edit. The clip you have moved is repositioned and all edits in front of it move forward in the Timeline.

You may notice that the clip has been inserted where you specified in the Timeline and that it also remains in its original position.

To remove the original clip, highlight it and press the delete key. This deletes the clip from the Timeline and leaves a gap where it previously existed.

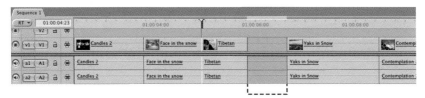

Gap Where Clip Has Been Deleted

To get rid of the gap hold down the Control key and click in the gap with your cursor – this opens a dialog box with several options – select Close Gap and the gap will disappear.

Control Click in the Gap to Reveal Contextual Menu

Fill with Slug

The Gap Has Now Been Closed

Note: you can actually delete a clip and close the gap at the same time by highlighting a clip – hold down the shift key and at the same time press delete.

Most Apple keyboards have an additional Delete key to the right of the standard Delete key. Press this key and the close gap function is performed with a single keystroke.

Selecting Multiple Items in the Timeline

You will now be well aware that a clip in the Timeline can be selected by clicking once with the Pointer tool. If you wish to select more than one clip hold down the Shift key while you highlight each of the clips. So long as you continue to hold down the Shift key you can then select as many clips as you wish. These can then be moved within the Timeline using the Overwrite or Insert method.

Another way to select multiple items is to use the Arrow tool. By clicking once and holding, this tool can be extended to reveal various options.

EDITING

Extended View of Arrow Tool

The Arrow tool is useful for deleting or copying large portions of the Timeline. Simply use the method described below to highlight those clips you wish to work with.

1   Click once to select the horizontal Arrow tool.

2   Your cursor now becomes a horizontal arrow. Use this horizontal arrow and click in the middle of the Timeline. All the clips forward of the arrow will now appear highlighted.

3   Press Delete and the highlighted section will disappear.

4   Press Apple 'Z' to undo the effect.

To copy the selected items repeat the procedure and this time instead of pressing Delete press Apple 'C' – or go to the Edit menu at the top of your screen and scroll down to Copy.

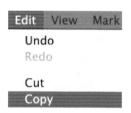

It is then possible to paste these items anywhere in the Timeline.

1   Select the Pointer tool.

2   Place the Scrubber Bar at the position in the Timeline where you wish to Insert or Overwrite the shots you have just copied.

3   Go to the Edit menu at the top of the screen. Select Paste to perform an Overwrite Edit on the section you have copied or select Paste Insert to perform an Insert Edit.

If you click the Toolbar and hold the horizontal arrow down with your mouse the Toolbar will extend to reveal other options within the arrow's capabilities. You can select a horizontal arrow which points backwards rather than forwards and thus select the contents of a track in the reverse direction; you can also select an arrow which points in both directions, allowing you to quickly and easily

77

select the entire contents of an individual track or everything in the Timeline (providing Linked Selection is switched on). You can also select a double arrow, forwards or backwards, which has the effect of selecting all tracks in the direction of the arrows regardless of whether Linked Selection is switched on or off.

 Selects Track/s in a Forward Direction

 Selects Track/s in a Reverse Direction

 Selects Track/s in Both Directions

 Selects all Tracks in a Forward Direction Regardless of Whether Linked Selection is Switched On or Off

 Selects all Tracks in a Reverse Direction Regardless of Whether Linked Selection is Switched On or Off

Cut, Copy, Paste

When using Final Cut Pro shots can be cut, copied or pasted using the conventions used in most word processors. These functions can be accessed from the Edit menu at the top of the screen.

To copy or paste a section from the Timeline is easy:

1. Highlight one or more clips in the Timeline and select Copy or Cut using the Edit menu at the top of your screen – or use the shortcuts Apple X (Cut) or Apple C (Copy).

2. Go to the Edit menu at the top of your screen and select either Paste (Apple V) to perform an Overwrite Edit, or Paste Insert (Shift V) to perform an Insert Edit of the copied or cut material.

Paste Insert (Shift V) is an extremely useful function, not found in word processors. As described above, to perform an Insert Edit select Paste Insert, while to perform an Overwrite Edit use Paste.

Snapping and Skipping between Shots

It is easy to skip between shots in the Timeline by dragging the Scrubber Bar which sticks to each of the edit points. If you have a crowded Timeline you may wish to turn this facility off as it can make it difficult to position the Scrubber Bar with accuracy.

Pressing the letter 'N' and snapping toggles on or off. You can also select the green symbol at the extreme top right of your Timeline to achieve the same result.

You can also skip between shots by using the vertical or horizontal arrows on your keyboard – up for forwards, down for backwards. Each press of these arrows will skip past one clip at a time.

The Razorblade Tool

My favorite tool in Final Cut Pro is the Razorblade. This tool is used for cutting clips into smaller pieces and is great for trimming a long shot into a smaller shot or shots.

It is often useful to use the Razorblade in conjunction with the Magnifier tool. By using the Magnifier you can zoom in on a clip or series of clips for greater accuracy when trimming with the Razorblade.

1 Play your Sequence in the Timeline. When you see a shot you would like to trim press the Space Bar to pause playback.

2 Click on the Razorblade tool – your cursor now becomes a Razorblade. Alternatively, press the letter 'B' to select the Razorblade tool.

3 Position the Razorblade where you wish to cut the shot in the Timeline. The Razorblade will automatically be drawn to each edit point providing you have Snapping switched on. If you find the Snapping facility to be impeding your accuracy then switch it off (press the letter 'N'). The Razorblade will also be drawn to the Scrubber Bar. It is therefore useful to position the Scrubber Bar at the exact point where you wish to perform a cut and then position the Razorblade accordingly.

4 Click once to make a cut at the point where the Razorblade is positioned. The track to which the cut applies is determined by whether or not the tracks are linked.

5 Select the Pointer tool and highlight the portion of the shot you wish to remove.

6 Press Shift Delete or press the Delete key to the right of the main keyboard area. This will remove the shot you have highlighted and close the resulting gap at the same time. If you wish to leave a gap only press Delete.

Should you happen to have Linked Selection switched off then the Razorblade will only cut through a single track at a time.

 If you wish to cut through all tracks, regardless of whether Linked Selection is on or off, then you need to select the Double Razorblade. This is done by holding your mouse button and clicking on the Razorblade tool – scroll across to select the Double Razorblade. The Double Razorblade will cut through all your tracks even if Linked Selection is switched off. The Double Razorblade can also be selected by pressing the letter 'B' twice in quick succession.

If you wish to heal a cut made by the Razorblade it is possible to perform what is called a Join Through Edit.

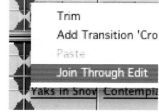

1 Control click on the cut made by the Razorblade. This will bring up a contextual menu.

2 Select Join Through Edit and release your mouse button. The cut will now be healed.

Note: The Razorblade does not cut through tracks which are locked. You need to switch off the locks for the Razorblade to work.

The Magnifier Tool

If you find it hard to be accurate when positioning the Scrubber Bar or Razorblade then you need to expand the Timeline. This is achieved by pulling on the ribbed ends of the Slider tool at the bottom of the Timeline, or by using the Magnifier tool.

The main difference between the Magnifier tool and the Slider tool is that the Magnifier tool is used to zoom in on a specific section of the Timeline. By using this tool the exact area one wants to magnify will get larger with each press of the button. When using the Slider tool the overall spread of the Timeline is increased or reduced, but not necessarily the specific area you wish to focus in on. The Magnifier is far more accurate.

1 Click once on the Magnifier tool in the Toolbar – your cursor becomes a magnifying glass.

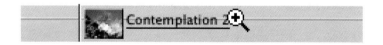

2 Position the magnifying glass over the area of the Timeline you wish to enlarge. Click with the mouse and the Timeline will expand – click again and it expands further. You can continue expanding the Timeline until you are able to work with the individual frames.

3 To contract the Timeline when using the Magnifier press the Alt key and a minus symbol will appear, indicating that the

Magnifier will contract the Timeline. Alternatively, click on the Magnifier in the Toolbar and select the minus Magnifier.

The Magnifier tool can also be selected by pressing the letter 'Z'.

Bringing Clips Back into Sync

Due to the nature of Insert Editing and Overwrite Editing, along with the fact that you can lock/unlock your tracks or activate/deactivate tracks through the break-off tabs, it is inevitable that at some stage your video and audio will get out of sync. Fortunately, Apple has made it very easy to bring items back into sync.

If items are pushed out of sync a red box will appear at the beginning of the clip in the Timeline where the sync trouble has occurred. The red box, which appears in both the video and audio tracks, will have a figure indicating the amount of sync slippage.

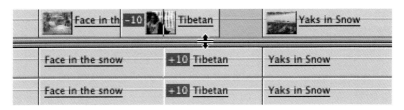

To bring the items back into sync:

1   Hold down the Control key and click inside the red box where the numeric value of sync drift is displayed. Control clicking in the red box brings up a dialog box with two possible options.

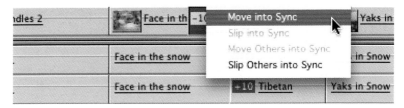

2   Select Slip into Sync or Move into Sync and release the mouse button. The item or items selected will now be brought back into sync.

It is also possible to manually slide the audio or video back into sync.

1 Make sure Linked Selection is turned off and Snapping is turned on. Having Snapping turned on will ensure that the out of sync items will be drawn to the correct sync points.

2 Highlight the item/s you wish to move and drag these so that the out of sync items line up at the beginning of each clip. When you release the mouse button the red box disappears and sync has been restored. If the red box is still visible move the clip again until you find the sync point.

Creating New Sequences

Final Cut Pro is particularly flexible in that you can have many Sequences open at a time. To have multiple Sequences open means that you have access to more than one Timeline – this is most useful for building various sections of a film which can be later joined together using the Cut, Copy, Paste and the Paste Insert functions. Think of having multiple Sequences as like having several different film reels each containing separate edited scenes or parts of a movie.

Multiple Sequences are tiled with cascading tabs from left to right. These tabs are displayed in both the Timeline and Canvas. Click on any of the tabs to flip between the Sequences. The Sequences can be renamed by overtyping the name in the Browser. The label on each of the tabs will then be updated.

THE FOCAL EASY GUIDE TO FINAL CUT PRO 4

1 To create a new Sequence click once inside the Browser. Hold the Apple key and press the letter 'N'. A new

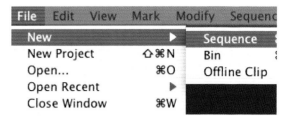

Sequence will then appear. Another way to create a new Sequence is to select the File menu (top left of screen). Scroll down to New Sequence and release the mouse button. A new Sequence will appear in the Browser.

2 Double click on any Sequence in the Browser to open it. Several Sequences can be opened at any one time. It is possible to rename Sequences at any stage by highlighting the Sequence icon in the Browser and overtyping.

Subclips

One of the easiest and most useful functions available in Final Cut Pro is the Make Subclip command. A subclip is part of a larger clip – it is therefore possible to have a long clip and to break this clip into many smaller pieces which can be individually named.

1 Open a clip into the Viewer.

2 Mark an 'in' and 'out' point – this will define the beginning and end of the subclip.

3 Go to the Modify menu – select the Make Subclip command.

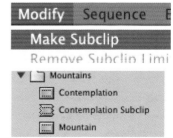

4 In the Browser a new icon will appear – the name will be that of the clip you have opened in the Viewer with the word Subclip after the title.

84

EDITING

Once you have your subclips in the Browser you can organize them into bins in the same way as you would with clips. The subclips can then be edited into the Timeline and worked with in the same way as clips.

Freeze Frame

When working with NTSC video each second is made up of 29.97 individual frames; when working in PAL each second is made up of 25 frames. It is a simple procedure to freeze any of these frames and create what is known as a Freeze Frame.

1 Position the Scrubber Bar on the frame you wish to freeze in either the Timeline or the Viewer.

2 Select the Modify menu at the top of the screen and scroll down to Make Freeze Frame.

Modify	Sequence	Effects
Make Subclip		⌘U
Remove Subclip Limits		
Make Independent Clip		
Make Freeze Frame		⇧N

3 Release your mouse button and the Freeze Frame with 'in' and 'out' points marked is positioned in the Viewer. By default a 12 second freeze is created (this can be set in User Preferences found under the Final Cut Pro menu).

4 If you want the Freeze Frame to be accessible within the Browser then drag the frame from the Viewer into the first column of the Browser. The Freeze Frame is represented by a graphic symbol within the Browser when viewed in list mode.

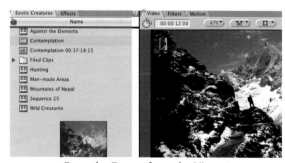

Drag the Freeze from the Viewer

85

If you find a Freeze Frame to be jittery you need to add the deinterlace filter. An explanation of using filters is described later in the Effects section.

Match Frame Editing

In on-line edit suites it is often necessary to perform a function known as Match Frame Editing. This means a specific frame is cued on a record machine, and an identical frame is cued on a source machine. By editing from the source VTR to the record VTR a seamless edit is performed. This facility was particularly useful in linear, multi-machine edit suites, before non-linear technology existed, when the editor did their best to minimize dubbing shots from one tape to another.

Match Frame Button

Match Frame Editing is still relevant in non-linear edit suites, such as Final Cut Pro, although it can be used for different reasons. It can be extremely useful to be able to find an exact frame in a Sequence with absolute accuracy. This technique can be used to locate a clip quickly and to put this clip into the Viewer for easy access.

To achieve a Match Frame Edit:

1 Place the Scrubber Bar in the Timeline on the frame you wish to match to in the Viewer.

2 Press the 'F' key on the keyboard or press the Match Frame button in the Canvas.

3 The frame on which you are positioned in the Timeline will now be displayed in both the Viewer and the Canvas.

4 By marking the 'in' and 'out' points in the Viewer and (if required) the Timeline, you can perform a seamless frame accurate edit. This is what is known as Match Frame Editing.

Note: if you are working with many layers of video or many audio tracks, and you wish to match frame to a particular layer or track, then you need to use the Auto-Select Toggle facility. This is located to the right of the locks in each track in the Timeline.

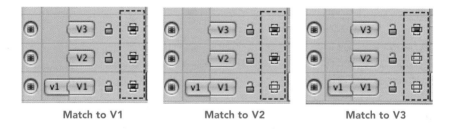

Match to V1   Match to V2   Match to V3

The Match Frame function, by default, will match to the base layer. Switch off the Auto-Select on the base layer and the Match Frame facility will then match to the second layer of video. Switch off the second layer of video and you can then match to the third layers of video, and so forth.

The same applies to the audio tracks in descending order.

While it may not be immediately clear exactly how useful Match Frame Editing is, there are many instances where it can make the difference between being able to successfully create an effect or not. It is also the easiest method of finding a shot without having to look through the Browser and all of the bins. Simply line up the Scrubber Bar in the Timeline – press the 'F' key – and the shot is immediately matched to.

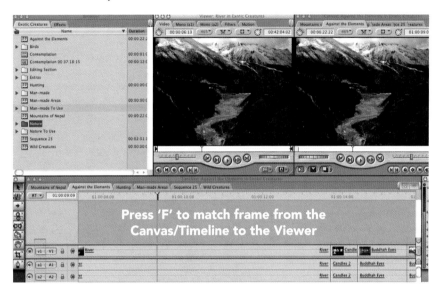

The Positioning of the Scrubber Bar in the Timeline is Matched to in the Viewer

Slow/Fast Motion

No editing program would be complete without being able to perform slow motion or fast motion to a clip or series of clips. Final Cut Pro performs admirably in this area, giving the freedom to slow images down to 1% or to speed images up in excess of 1000%. It is also possible to play images in reverse.

1   Click once in the Timeline and highlight the shot you wish to slow down or speed up.

2   Go to the Modify menu at the top of the screen and scroll down to Speed.

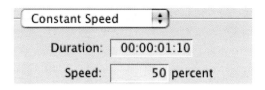

3 Enter a percentage value to set the speed you wish the shot to play at. If you want to play the clip at half speed enter 50%. If you want the shot to play at double speed enter 200%.

4 The duration of the shot in the Timeline will now change according to the value entered. Depending on which Mac you are using the shot may now have to be rendered for playback. If there is a red line above the shot this indicates the shot must be rendered. Other colors indicate some level of playback will be possible.

5 Select the Sequence menu at the top of your screen and scroll down to Render Video. Check that there is a tick next to the color which corresponds with the colored line above the shot in your sequence. For example, if there is a red line you will need a tick next to the red Needs Render setting. Likewise, if there is a green line, check there is a tick next to the green Preview. These settings toggle on and off.

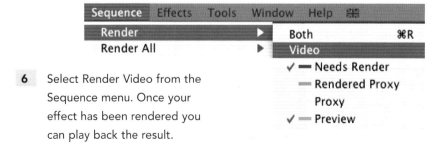

6 Select Render Video from the Sequence menu. Once your effect has been rendered you can play back the result.

To play a clip in reverse simply check the 'reverse' option in the speed dialog box or enter a negative value.

It is worth noting that you may be able to play back the result without rendering by using the RT Extreme facility within Final Cut Pro. For a true representation of the final result it is best to render the clip. This will give you full resolution playback.

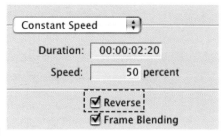

Split Edits

In most situations where an edit takes place both video and audio will cut at the same point. This is fine for most situations. However, in other situations a technique known as a Split Edit may be used. A Split Edit is where audio and video are not cut at exactly the same place – either one may precede the other and this can apply to either the 'in' or 'out' point of an edit – thus audio or video may start and/or finish at separate points.

Split Editing is a technique often used in news and documentaries, particularly in interview situations. For example, you may hear a person speaking over visuals of a scene being described. After several seconds, with the voice of the person still running, the video will cut to the person speaking.

To achieve a Split Edit:

1   Open a shot into the Viewer.

2   Position the Scrubber Bar where you wish to mark the edit.

3   Hold down the Control key and click in the white area in the Viewer (where you mark 'in' and 'out' points). This will open the contextual menu which is used for

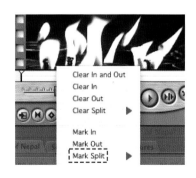

clearing and setting 'in' and 'out' points. At the bottom of this menu is an option for Mark Split.

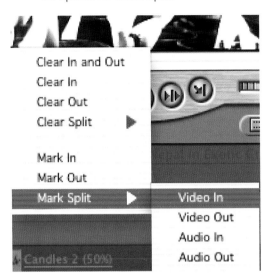

4 Scroll to Mark Split and this opens a menu to the right giving you options such as: Video In, Audio In, Video Out and Audio Out.

5 Choose an option, whether it be Video In or Audio In and release the mouse button.

6 Reposition the Scrubber Bar and now choose the opposite choice – if you have already marked Video In, now choose Audio In.

7 Repeat the process for the end of the edit. You can mark separate video and audio 'out' points. If you do not want separate video and audio 'out' points then mark an 'out' point by simply pressing 'o'.

8 Position the Scrubber Bar in the Timeline and use Overwrite to edit the shot into the Timeline.

Play back the edit and watch the result. If you have followed all of the above steps the audio and video will cut in the Timeline at separate places. If you marked a split for the end point audio and video will finish at separate places.

Achieving Split Edits in Final Cut Pro is not the easiest of functions but it is definitely worth learning. Once mastered you can add finesse to a film and boost it into the professional realm. Any editor worth their salt will understand the process of performing Split Edits and know how and where to use them.

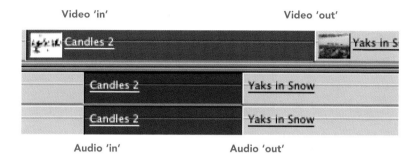

The Split Edit is also reflected visually in the Timeline. If you look at your audio and video tracks you will be able to see where the audio and video edits take place. If it is not clear you may have to use the Magnifier tool to increase the size of the Timeline and then each of the edits will be more obvious.

If you find the above process difficult to follow there is another way to achieve Split Edits. This is done by Locking Tracks and editing directly in the Timeline with the Razorblade.

1   Lock the tracks you do not wish to affect, for example lock your two audio tracks.

2   Select the Razorblade tool from the Toolbar.

3   Cut the video at the point where you want the Split Edit to occur.

4   Choose the Pointer tool and highlight the piece of video you wish to remove. Press the Delete key.

5   Drag the end of the shot to fill the gap created when the razorbladed section was deleted.

You may find that it is not possible to drag the end of the shot to fill the gap. This could be because you are at the end of the media limit. All non-linear editing systems work by referencing to the original media stored on hard disk. If all the media of a particular clip is already edited into the Timeline then it will be impossible to extract more frames.

Another way to achieve Split Edits is to use the break-off tabs in the Timeline and to restrict the flow of video or audio. For example, if the break-off tabs for Audio 1 and Audio 2 are disconnected then only the video will flow through when performing an Overwrite Edit. Using the break-off tabs is like turning a tap on or off. If it's connected it is open – if it's disconnected it is closed.

Drag and Drop Editing

The editing we have done so far has involved opening a clip in the Viewer, marking an 'in' and 'out' point, and then editing this clip into the Timeline via the Canvas. It is possible to bypass this method altogether and edit clips directly from the Browser into the Timeline, or to edit from the Viewer to the Timeline without involving the Canvas in the equation.

1. Create a New Sequence in the Browser (press Apple N).

2. Double click the new Sequence to open it – you can rename the Sequence if you wish.

3. In the Browser click once on any of the clips – do not release the mouse button (open a bin if all your clips are filed away).

4. With the mouse button still depressed drag the clip from the Browser directly into the Timeline. Release the mouse button and your clip will be edited into the Timeline at the position where you released your mouse button.

5. Do this again with a few more clips and you will see that it is possible to build a sequence simply by dragging clips from the Browser to the Timeline.

6. Once you have several clips positioned in the Timeline repeat the procedure, however, this time drag the clip into the middle of the Timeline. Don't release your mouse button just yet!

When you have your clip positioned in the Timeline move the mouse button gently and notice that if you have the cursor pointing to the top third of the video track there is a horizontal arrow – if you point the cursor towards the bottom half of the screen there is a vertical arrow. A horizontal arrow represents Insert Edit while a vertical arrow represents Overwrite Edit.

Drag Clip to Bottom of
Video Track – Overwrite Edit

Drag Clip to Top of
Video Track – Insert Edit

The biggest disadvantage with using Drag and Drop editing is that you do not have the control over marking the 'in' and 'out' points of the clip in the Viewer. However, this can be a very quick way to throw clips together into the Timeline. Furthermore, you can also do two other very useful tricks using Drag and Drop.

First, if you work in picture icon view you can arrange the icons in whichever order you choose (arrange them left to right in storyboard fashion) and then, by selecting an entire group of clips (by lassoing, or by

using Alt and clicking to select multiple clips) you can then drag as many items as you wish into the Timeline. These clips will be positioned in the Timeline in order of the icon arrangement.

Buddah Impressive	Candles 3	Elephant
dah Impressive	Candles 3	Elephant
dah Impressive	Candles 3	Elephant

Once the clips have been dropped into the Timeline the Razorblade can be used to chop away unwanted sections. If you really wanted, you could edit in this way without ever opening a clip in the Viewer or dragging across to the Canvas (however, I would never rely on this exclusively as my method of editing).

Highlight the Icons in the Browser and Drag these Directly into the Timeline

The Clips will then be Arranged in the Timeline in the Order they were Positioned in the Browser

It is also worth noting that it is possible to drag clips from the Viewer to the Timeline, thus skipping out the step of editing across to the Canvas. The same rules apply when dragging clips from the Browser to the Timeline. If you have a horizontal arrow this will represent an Insert Edit, while if you have a vertical arrow an Overwrite Edit will occur. You can also mark 'in' and 'out' points in the Viewer in the usual way. These 'in' and 'out' points will apply and therefore determine the beginning and end of the edit.

If clips are not open in the Viewer the 'in' and 'out' points will still apply. Thus should you drag a clip directly from the Browser into the Timeline, the duration of the edit and the start/end frame of the clip will be defined by the 'in' and 'out' points which have previously been set.

Extending/Reducing Clips by Dragging

Clips can be made longer or shorter by grabbing hold of either end and dragging the length in either direction.

1. Choose a clip in the Timeline which you wish to extend or reduce. Position the pointer and let it hover over the center of an edit point or the area where two clips meet. A symbol with two vertical lines and two horizontal arrows will appear.

2. If you wish to reduce the length of the clip drag the end of the clip into itself (using the symbol with two vertical lines). A display will appear to the right showing the overall clip duration and the trim adjustment in seconds and frames.

At the same time the Canvas will shuttle through the clip as you drag the end, giving you a visual reference to the adjustments being made. You therefore have a display in both the Timeline and the Canvas giving you numeric and visual indicators at the same time.

Providing you have Linked Selection switched on audio and video will move together, otherwise they will be independent.

3. In order to extend the length of a clip a gap must exist between the clip you wish to extend and the clip adjacent to it. To create a gap

either insert a shot and then delete it and a gap will be created, or use the Arrow tool to highlight and drag several clips further along the Timeline.

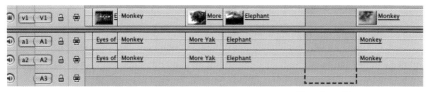

Create a Gap

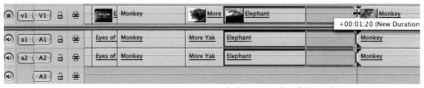

Extend the Length of the Clip into the
Gap you have Created

If you find it impossible to extend the length of a clip it most likely means there is no more material left to extract from the original clip. All clips have a media limit as defined by the amount of material originally captured to hard drive. Once this media limit is reached you can go no further.

For fine control when dragging, hold down the Alt/Option key. Another way to achieve fine control is to expand the Timeline by using either the Magnifier tool or the Slider Bar at the bottom of the Timeline.

RENDERING

> *It is a unique editor. And there's nothing like it out there.*
> CHARLES ROBERTS
> MEDIA PROFESSOR, AUTHOR, EDITOR

Rendering is the process by which your computer builds each of the individual frames needed to produce an effect. When you play back straight cuts in the Timeline nothing needs to be rendered. The computer simply refers to the hard drive where the original shot has been recorded and uses your edit information to determine which section of the original shot is needed. When an effect is applied to a clip a different process must take place.

People often complain about rendering – to wait seconds or minutes for a computer to produce an effect can often drive people mad. I always smile at these situations. My background was in the world of on-line tape editing where the editor would work with several videotape machines, a separate vision mixer, character generator, audio mixer and Digital Video Effects (DVE) generator. To produce effects in this setup would often require an editor to record one or more of the shots to separate tapes. Several tape machines would then be run in sync with the layered effects built through the mix-effects banks of a vision mixer. The output would be in real time, however, real time was only achieved at the expense of the time used in the setup of an effect. These effects would often take a considerable amount of time to set up.

It is not always necessary to render effects in Final Cut Pro 4 to see a result. Through the addition of a facility known as RT Extreme it is now possible to see many effects play back in real time directly through the Firewire. One should be aware, however, that the output quality is not full quality and that the results are intended for the purpose of gauging timing and other critical decisions, rather than for producing full resolution output. The final output will still need to be rendered. The advantage of RT Extreme is that it allows the editor, in many situations, to build effects without having the render process getting in the way of one's creative flow.

Note: in order to access RT Extreme the minimum requirements are a G4 500 MHz computer with 1 MB level 3 cache and a minimum of 512 MB of ram. In other words, you will need a powerful Mac with loads of ram. The more the better.

The amount of real time one gets through RT Extreme depends on the processor inside your Mac, whether you are using a G4 or a G5 machine, the amount of installed memory, the system bus speed of the computer, how many layers, filters, transitions and generators are being used and the complexity of the effects created.

A further factor to consider is the setting chosen in the Real-time Effects pop-up menu. The output quality for RT Extreme is customizable. You choose the quality you want the output to be played back at.

Unlimited RT allows Final Cut Pro to play the maximum possible in the way of real-time effects, however, the trade-off is an increased likelihood of dropped frames. Alternatively, one can choose **Safe RT**, which will play the result without dropped frames, however, this will limit the overall ability to

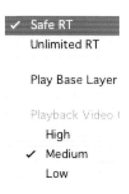

play back effects in real time. Furthermore, one can choose to vary the output quality between high, medium and low. These settings directly affect the output resolution of the video signal.

My choice is to leave the RT Extreme settings at Unlimited RT and vary the output quality from low to medium to high. I can live with dropped frames on playback and I can also live with low-resolution playback, if need be.

The idea of RT Extreme in my opinion is to get a feel for how an effect will look. For a true representation one must render for final output. There is no other way to know exactly what the final result will look like.

The Render Settings

In previous versions of Final Cut Pro the render settings were basic. One could choose to render everything in the Timeline or one could choose to render a particular section. Now the choices are far greater. The choices give power to the editor. One can be quite specific about which sections of the Timeline one wishes to render.

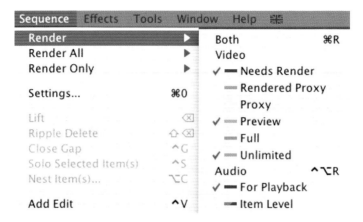

1. Choose the Sequence menu and scroll to Render – a window opens displaying various colors. These are the options one can choose when setting the level to which you want the Render command to apply.

2. You can choose as many or as few of the colors to apply to this setting as you wish. Each level, represented by a color, works by means of a toggle. A tick means a color is selected; no tick means it is not selected. You need to choose the colors you want to apply to the Render Selection setting. This is achieved by clicking on a setting and releasing your mouse button. The color that you clicked will now have a tick next to it. Each time you wish to activate or deactivate a particular setting you need to repeat the process.

3. Do the same for the Render All settings under the Sequence menu and the Render Only setting. Render All applies to the entire Timeline, whereas Render Only applies to a particular color, or level of render,

RENDERING

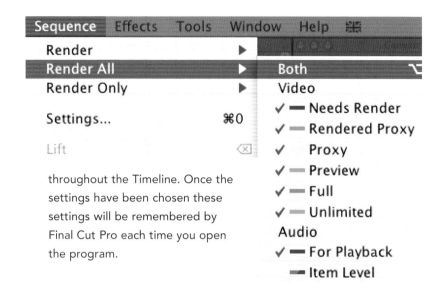

throughout the Timeline. Once the settings have been chosen these settings will be remembered by Final Cut Pro each time you open the program.

At a basic level, one could select the color red. When a red bar is exhibited above a clip, it means the real-time capabilities of Final Cut Pro have been exceeded and the material must then be rendered for playback.

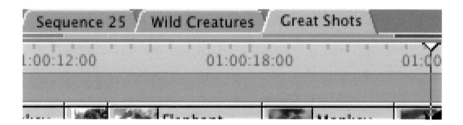

The following is a basic summary of what the most important of the individual colors mean:

Dark Gray – no rendering is required.

Steel Gray – material in the Timeline has been rendered.

Dark Green – will play back through RT Extreme at full quality.

Green – will play back through RT Extreme, however, motion and scaling effects will be approximate only.

Yellow – will play back through RT Extreme with an approximate representation of the effect. Some filter controls may be ignored.

Orange – effects will be played back with a high likelihood of dropped frames.

Red – requires rendering for playback.

The render commands I use most are to Render a particular part of the Timeline – accessed by holding down the Apple key and the letter 'R', or Render All, accessed by holding down the Alt/Option key and pressing 'R'. This will render all the video in the Timeline that requires rendering. By accessing the Sequence menu the commands can also be executed. One can choose to render Both – which means Video and Audio, or Video or Audio Independently.

Apple + R – Render highlighted video in the Timeline
Alt/Option + R – Render all video in the Timeline

You need to configure the render options to suit your needs. My choice is to keep life really simple. I select all the options except for the Item Level choice in audio. This means I know that everything necessary for a full quality render will be performed when I hit any of the render commands.

Whenever you see a red bar at the top of the Timeline (above a shot to which an effect has been applied) this means this section must be rendered. It is possible to render individual clips or to instruct the computer to render everything in the Timeline that needs rendering.

When a shot requires rendering, go to the Sequence menu at the top of your screen and scroll down to Render. Providing you have the appropriate colors selected to match the color of what needs to be rendered, the process will then begin.

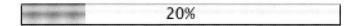

A progress bar will display in percentage terms how much material has been rendered.

Once the render is complete you can then play the rendered clip(s) in real time. If you are happy with the result, carry on editing. If you are not happy make the necessary adjustments to the parameters of the effect and then render again.

At any time you can stop a render in mid-progress by pressing the Escape key (located top left of your keyboard). The render will cease, however, the portion of the shot or Sequence already rendered can be played back. This is a particularly useful feature as one can choose to render a small portion of a shot or Sequence, then play back this section to determine whether or not to go ahead with the complete render. Once you restart the render process begins again at the point where it was previously stopped. Thus, you do not need to re-render material that has already been rendered just because you stopped the process mid-render to look at the result of an effect.

A final useful work tip. If you mark the 'in' and 'out' points in the Timeline, you can then choose the Render In to Out command found under the Sequence menu. This will render everything between the 'in' and the 'out' points. Very useful!

MEDIA MANAGEMENT

> *You have high-end, professional, 'broadcast and beyond' capable tools in an easy to use package that anybody can learn.*
> JIM KANTER
> ATLANTA FINAL CUT PRO USER GROUP

In an ideal world when one completes a project it should be as simple as deleting the files from the scratch disks which were chosen during initial setup. In reality, editors will often end up with files scattered across several hard disks with everything from clips to render files in the most obscure and hidden areas of one's computer. This situation is particularly common when one works with several projects at a time and when one fails to pay attention to the setup of scratch disks when moving from project to project. Fortunately there are a few very easy techniques which one can use without having to search through every folder on each hard drive to track down all of the files that have been used in a particular project.

Making Clips Offline

An invaluable command is **Make Offline** found under the Modify menu. Make Offline, as the name suggests, will break the link between the project you are working with and the clips you choose to make offline. This is done in one of three ways:

(i) the link is broken and the clips are left stored on the hard drive

(ii) the selected clips are moved to the trash

(iii) the selected clips are deleted from the disk.

MEDIA MANAGEMENT

1   Highlight the items in the Browser which you wish to Make Offline. Clips, bins or Sequences can be chosen.

2   Select the Modify menu and scroll to Make Offline. A box will open giving you three options.

3   Choose the option you wish to perform.

After making selected clips offline...

Action For Media Files:
● Leave Them on the Disk
○ Move Them to the Trash
○ Delete Them from the Disk

You can choose to make the clips offline and leave them stored on the hard drive where they already exist, or move them to the trash, or delete them from disk. If you choose either of the last two options be careful as the results cannot be undone.

I often use Make Offline and select Delete Them from the Disk at the end of a project. This ensures a quick easy way to clear the captured media files from the hard drives in your computer with the minimum of effort on your part.

It can also be useful to move a bunch of files, or all the files of a project, to the trash, to copy elsewhere. This is the simplest way of tracking down all the files in a project without having to manually search for them.

The Render Manager

Whenever you render material in the Timeline the files created during the render process are stored in the Render folder/s created when you set your scratch disks. Due to the nature of building effects material will often be rendered several times over until you get the effect exactly as you want it. It can therefore be advantageous to get rid of old render files which are not being used anymore and reclaim the hard drive space. This is exactly what the Render Manager does for you – it clears out the render files you no longer need in a simple and efficient way.

1   Choose the Tools menu, scroll to Render Manager and release your mouse button. A window will open displaying folders which reference to the render files on your hard drives.

2   Click the downward arrows to reveal the render files. Information to the right will tell you the amount of hard drive space these files take up.

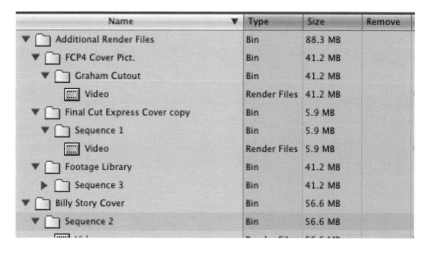

3   Highlight the files you wish to remove. Apple click to highlight multiple files, one at a time, or Shift click to highlight everything from top to bottom.

MEDIA MANAGEMENT

Name	Type	Size
▼ 📁 Additional Render Files	Bin	88.3 M
▼ 📁 FCP4 Cover Pict.	Bin	41.2 M
▼ 📁 Graham Cutout	Bin	41.2 M
🎞 Video	Render Files	41.2 M
▶ 📁 Final Cut Express Cover copy	Bin	5.9 MB
▶ 📁 Footage Library	Bin	41.2 M
▼ 📁 Billy Story Cover	Bin	56.6 M
▼ 📁 Sequence 2	Bin	56.6 M
🎞 Video	Render Files	56.6 M

4   Click OK and these files will be deleted from the disk.

Note: the results cannot be undone as the files have now been removed from disk, however, you can simply re-render the material if you find you have made a mistake. Nothing will be lost but the time it takes to render.

> *There are no special tools anymore.*
> *Everybody has access to the same stuff.*
> JOE MALLER
> JOE'S FILTERS

When it comes to building effects in Final Cut Pro there are several things that need to be considered. First, there are the type of effects that are applied to individual clips or to a series of clips in the Timeline – these are single layer video effects. Then there are the types of effects that are termed multi-layered effects.

Look to the video track symbols at the front of the Timeline – the break-off tabs are labelled V1, A1, A2…. These are the video and audio tracks. Final Cut Pro allows for up to 99 video tracks or layers to be created.

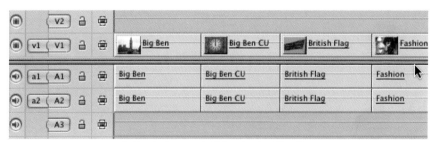

A Single Layer of Video in the Timeline

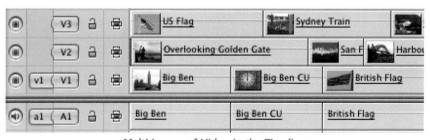

Multi-Layers of Video in the Timeline

Creating a video track – which means adding a layer – is as simple as Control clicking in the gray area next to any of the tracks which already exist.

A menu will open giving you the choice to either add or delete a track. You can Control click above, below or to the left of the tracks in the Timeline to open up the contextual menu which allows you add or delete tracks.

Control Click in this Area

When working with a single layer of video, effects can be applied in the form of Transitions, Filters and Generators.

Transitions refer to an effect that is applied between two clips. Examples include dissolves, wipes and slides.

Filters can be applied to a clip or part of any clip. They are used to change the look of images by manipulating each of the video frames which make up the image. Examples include blurs, mattes, borders and changes to brightness, contrast and color.

Generators are neither Transitions nor Filters. These are used to create additional material which you need to work with. These include devices such as slug (black), text and matte colors. Think of Generators as being the electronic equivalent of spacer, head and tail leaders and countdowns of the film world.

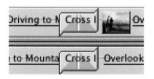

A Transition is Added Between Two Clips

A Filter is Added to a Clip

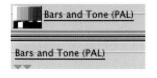

A Generator is Created Inside of Final Cut Pro

When working with Transitions and Filters these are applied to a single layer of video and can be integrated into a Sequence which is made up of no more than a single video track. This does not mean that Sequences made up of multiple layers do not use Transitions and Filters. The point is a Sequence made up of no more than one layer can include Transitions, Filters and Generators. Multi-layered Sequences will also make use of Transitions, Filters and Generators and these can and likely will be applied to any of the layers in the Sequence.

The instant you have more than one layer of video you are working in the realms of compositing. Layers are video components which are stacked on top of each other – examples include: picture in picture, titles, and animated components.

An Example of Compositing Made Up of Two Video Layers

An Example of Compositing Made Up of Three Video Layers

Before we move onto Compositing it is important to know how to work with Transitions, Filters and Generators. You need to understand the difference between these. Just to refresh your memory: Transitions work between clips; Filters are applied to clips; and Generators work in a similar way to clips but are generated within Final Cut Pro itself.

The Concept of Media Limit (Handles)

It is vital to understand when using Transitions that there must be available media for the Transition to work. This available media is referred to as 'handles'. The maximum length that a Transition can be is equal to the available media or 'handles' which exist on the original clips as they were captured to your computer's hard drive.

Thus, should you wish to apply a one second dissolve between two shots then there must be at least 12 frames (PAL) or 15 frames (NTSC) of available media. The available media applies to the end (tail) of the outgoing clip, on one side of the transition, and the beginning or 'head' of the incoming clip. The idea is the same as checkerboarding two pieces of film in an optical printer or A/B rolling shots on separate machines in a linear tape suite.

If you do not have the available media then the length of a transition will be restricted to the media that is available. It is impossible to exceed these limits.

Applying Transitions

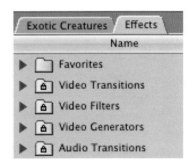

1. Click once on the Effects tab in the Browser and this will reveal a list of Transitions, Filters and Generators. I suggest working in list mode for this part of the operation.

2   Click once on the triangle to the left of the Video Transitions folder – this will reveal a list of the available Transitions. Choose the type of transition you require and click the triangle to the left. Now choose a transition you want to work with.

3   Drag the Transition onto the edit point in the Timeline where you want the effect to be applied.

4   A render bar will appear indicating the state of the material in the Timeline. If it is red, it needs to be rendered for playback, if it is green the effect will play in real time through RT Extreme. If it is yellow or orange the effect will play, however, the reliability of playback cannot be guaranteed.

Note: when choosing Transitions, Filters or Generators if the effect is bold under the Effects tab in the Browser, this means the effect will play in real time through RT Extreme. If the effect is not in bold then it will have to be rendered for playback.

Changing Transition Durations

1   Double click on the Transition in the Timeline – this will open a set of controls in the Viewer.

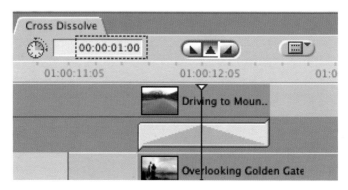

2 At the top of the Viewer is the name of the Transition. Immediately below this is a numerical value representing the Transition duration. Click once with the mouse in this box and this will highlight the numeric value. You can now overtype this value. If you want a 2 second dissolve type 200 and press the return button; if you want a half second dissolve type 12 (PAL) or 15 (NTSC) and press return; for a 10 second dissolve type 1000 and press the return button.

The Change in Transition Duration is Reflected in the Timeline

3 The dissolve duration is now changed. The above set of instructions applies regardless of which transition is chosen.

Applying Filters

Filters are extremely powerful and creative tools to work with. It can take a good deal of experimentation to master them. It is definitely worth the effort, however. What can be produced on a desktop editing system with Final Cut Pro would have cost a fortune at a facilities house only a few years ago. The price paid now is the amount of time the individual is prepared to spend building and rendering these effects.

1 Click on the triangle to the left of Video Filters in the effects area of the Browser. This will reveal a list of Filters stored in bins. Open a bin and choose a Filter you wish to experiment with.

2. Drag the Filter onto the clip where you want the Filter to be applied. Remember, if the Filter is bold this indicates it will play in real time through RT Extreme. If it is not bold it will need to be rendered. The level of real-time playback will be determined by the color of the render bar which appears above the clip in the Timeline.

3. To alter the parameters of the Filter double click the clip in the Timeline. Click on the Filters tab located at the top of the Viewer.

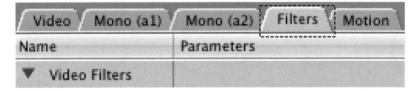

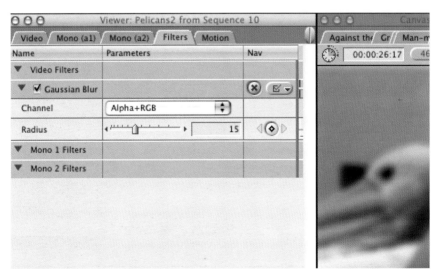

This will open a selection of controls that are used to alter the parameters of the filter.

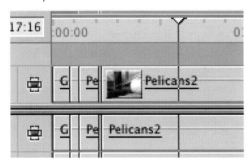

4 Make sure the Scrubber Bar in the Timeline is positioned on the clip to which the Filter has been applied. This is important as it ensures adjustments will be displayed visually in the Canvas.

5 Experiment with the controls under the Filters tab in the Viewer and observe the result in the Canvas and on your television screen (if you are set up with this configuration). Play the clip back through RT Extreme, if possible, or render the clip if necessary.

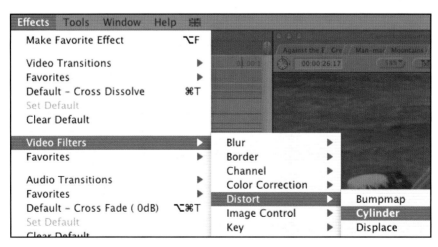

All Transitions and Filters can also be accessed from the Effects menu at the top of the screen. Simply scroll down to Video Transitions or Video Filters and select the effect you want. If you wish to use a Transition make sure the Scrubber Bar in the Timeline is positioned where the Transition is to be applied. If you choose a Filter, first highlight the clip in the Timeline and then choose the Filter from the Effects menu. Whether you choose to access the effects from the Effects menu or from the Effects tab in the Browser comes down to personal preference. The result will be the same.

Compositing

Compositing is where all the fancy stuff happens: flying titles, moving boxes, multi-layered dissolve sequences, transparent backgrounds – all the elements that 'dress up' a video production and make it more than cuts, dissolves, basic transitions and filters. As mentioned earlier, compositing encompasses everything that involves more than one layer of video.

Video filters, transitions and generators can all be applied to any video track, or several individual video tracks at a time, however, it is the 'stacking' or 'layering' of tracks of video that builds a composited sequence.

In on-line linear suites layering of video was done on a vision mixer in combination with a DVE (digital video effects). In the film world an optical printer was used.

Pieces of film were sandwiched together called bi-packing) and this was then exposed to several passes of light to achieve complex effects. Inside of Final Cut Pro the tracks are layered in hierarchical order with the tracks closest to the top having priority over those below.

For those who have no idea what a DVE is – this refers to a stand-alone box used in television production for creating special effects. DVEs became popular in the late 1970s and 1980s and have been used all the way through to the present. DVEs have been seen as providing the video equivalent to the optical printer of the film world. They have traditionally been regarded as horrendously expensive, powerful devices. They are still used in live television production and on-line edit suites, however, as a result of programs such as Final Cut Pro mere mortals can now achieve sophisticated effects with modest budgets. Previously a simple squeeze or flip (or flop as it is called in Final Cut Pro) required one to step into an expensive post-production suite and often pay far more than the cost of Final Cut Pro for a single effect!

Methods of Creating Multiple Tracks

By default Final Cut Pro opens with a single track of video in the Timeline. There are three possible ways to add more tracks:

(i) Control click next to the V1, V2 symbols (to the left of the locks). This will give you the option to either Add or Delete a track. If you click in the gray area above the last existing video track there will be a single option which is to add a track.

(ii) Drag a clip from the Browser or Viewer directly into the Timeline, into the gray area, above the top-most existing video track. Release your mouse button. This will then create a new video track where you drop the clip and two audio tracks below the last existing audio tracks.

(iii) Open the Sequence menu at the top of screen and choose the Insert Tracks command from the menu.

Once you have created the additional video tracks, items can then be edited to these tracks by dragging clips from the Browser or Viewer directly into the Timeline, or you can direct the flow of video/audio using the break-off tab patching facility in combination with Insert and Overwrite Editing.

The Motion Tab

The Motion tab is home to a great many of the tools used in the compositing process. In this window you will find the facility to resize images, rotate, reposition, crop, distort and adjust the opacity of clips. Effectively the Motion tab provides you, the editor, with a fully fledged DVE facility.

To access the Motion tab, click the last tab to the right in the Viewer. Click the arrows to the left of each of the headings to access the controls.

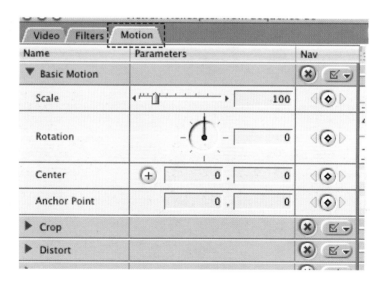

Using the Motion Tab

To **scale** an image means to reduce or increase its size. The default size is 100%, therefore numbers greater than 100 mean the image has been made larger or 'blown' up whereas numbers less than 100 mean the image has been reduced in size.

To alter the size of a clip using the Motion tab:

1   Double click a clip in the Timeline. This will open the clip into the Viewer. Make sure the yellow Scrubber Bar is positioned on the clip in the Timeline. This ensures you will see the result in the Canvas as you alter the parameters of the clip in the Motion tab.

2   Click the Motion tab in the Viewer.

3   Move the slider bar next to the word Scale either backwards or forwards. Alternatively type in a number. If you wish to reduce the clip in size by half type 50%. If you want to double its size type 200%.

Providing the clip is positioned on video 1 in the Timeline the result will be the image, reduced in size, over black. If you do not see this result you need to position the yellow Scrubber Bar over the clip in the Timeline prior to reducing the image in size.

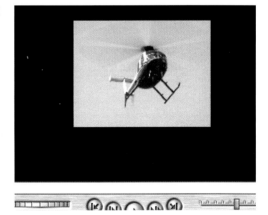

Now repeat the process, however, this time we want to work with two tracks of video.

1   Create a second track of video by Control clicking in the gray area above V1.

2 Select Add Track.

3 Slide the V1 break-off tab to V2 and edit a clip onto V2. Alternatively, you can drag a clip direct from the Browser or Viewer to V2.

4 Reduce the size of an image on the V2 track to 50%. This will create what is called a 'picture in picture'. The image on the V2 track will be reduced in size while that on V1 remains at 100%.

Providing your computer is capable of RT Extreme, and External Video is set to All Frames and the Firewire cable is connected to a deck/camera, then you will be able to see the result play in real time onto a television monitor. If you are working exclusively on the computer monitor and not connected to a deck/camera then you can observe the result in the Canvas.

To rotate the image on the V2 track simply select the rotate command underneath Scale.

Turn the wheel or type in a numeric value.

Remember – a circle is made up of 360 degrees.

▼ Basic Motion	
Scale	
Rotation	
Center	⊕ 0 .

Again, to ensure the clip is active in the Viewer, first double click it in the Timeline. Then click the Motion tab to see the controls, making sure the yellow

EFFECTS

Scrubber Bar is positioned on the clip in the Timeline/Canvas. This will ensure the result is displayed in the Canvas.

To move a clip about within the Canvas click on the plus '+' symbol to the right of the word Center in the motion controls. A small '+' will then appear in the Canvas (you might have to look hard to see it – but it is there!) Click the '+' with your mouse and reposition the image within the Canvas window. Alternatively, type co-ordinates into the boxes.

As you work, you may choose to rotate the image again or resize it. Other parameters can be altered. For example, check the Drop Shadow box towards the bottom of the Motion window.

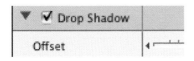

Experimentation is the key. By adjusting parameters, typing numbers, adjusting opacity, altering angles – many different results can be achieved. No instruction manual can ever teach you everything that is possible. It is up to you to teach yourself by altering the controls and trying out the possibilities.

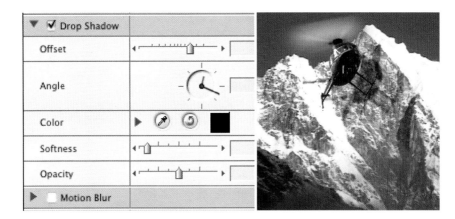

Image + Wireframe

An important and useful mode is what is called Image + Wireframe. Click the arrow in the box located towards the center of the Canvas. Select Image + Wireframe from the drop-down menu.

Once you have selected Image + Wireframe you will notice a large cross will appear from end to end across the active image. When working in this mode it is possible to slide the

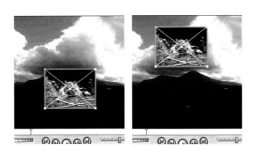

Click on Center of the Image and Reposition by Dragging

image around the frame by simply clicking on the center of the image with your cursor and then repositioning by dragging. This is a lot simpler than typing in center co-ordinates into the boxes or using the little '+' symbol as described earlier.

Below: The image can be rotated by positioning your cursor on either edge and moving your mouse in opposite directions.

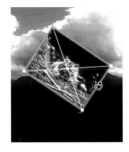

Your Cursor Becomes a Circular Arrow. You can then Rotate the Image

Below: By dragging any of the corners of the image it can quickly be resized.

Position Your Cursor on the Edge of the Image and Drag to Resize

Using Image + Wireframe provides a quick and effective way to reposition, resize and rotate images. The downside is you do not get the same control which you do by entering numerical values directly into the Motion Control window. Choose whichever option suits the task you need to perform.

Titlesafe

Under the same menu as Image + Wireframe is another setting called Titlesafe. Both settings, Image + Wireframe and Titlesafe, can be switched on or off in either the Viewer or Canvas.

If you find it impossible to switch on Titlesafe you need to first ensure that Show Overlays is checked. With Show Overlays selected it is then possible to toggle Titlesafe on and off.

The Titlesafe setting is particularly important to have switched on when positioning images or working with titles. Domestic monitors do not display the full video image as it is recorded to tape. What is displayed on your computer monitor known as 'underscan' while what you see on your television set is known as 'overscan'. Broadcast monitors offer both options.

When Titlesafe is switched on it will be obvious – two sets of blue lines will be noticeable around the inside perimeter of the Viewer or Canvas. The Title Safe area is known as the Essential Message Area (EMA). In simple terms, to ensure that the images you are working with will be seen correctly on a television set you need to make sure they are positioned within the lines of the Title Safe area. The outer lines are regarded as safe on most televisions, whereas the inner lines are regarded as safe on virtually all television sets. When working with text it is best to position the text inside the inner lines. When positioning images do not go outside of the outer lines unless you are comfortable that these images will likely spill outside the frame, or cut-off, on playback. Of course, allowing images to spill outside of cut-off can be used for effect, however, one needs to know that this will take place. In other words, if this effect takes place it should be deliberate, rather than accidental.

It is important to remember that not all television sets are the same. Some will crop the final images on playback more than others.

Working with Multi-Layers

It is relatively easy to create a picture in picture as already described. Using the same principles one can position several images on-screen at a time. The effect we will produce is to have a single image positioned as a background with three other images layered over it. Prior to the availability of systems such as Final Cut Pro this type of effect could only be produced at facility companies and television stations using powerful DVEs and other expensive equipment. Now you can do it on your desktop quickly and easily.

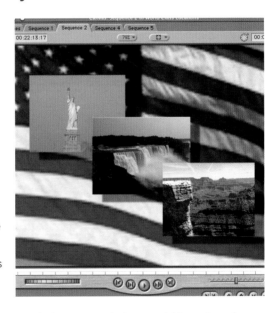

1. Layer the images you want to work with over each other in the Timeline. Be aware that the order in which they are stacked will determine the priority of the layers.

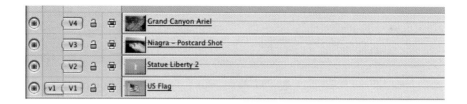

Thus the image on layer 4 will be at the front of the layers, while the image in layer 1 will serve as the background. Those on V2 and V3 will make up the middle layers.

2   Reduce each of the images in size to 35% – except for the background image, this stays at 100%. To achieve this you will need to individually click on each of the images to bring them into the Viewer, and then resize accordingly. To achieve the same size for each image make sure the number in the Scale setting is the same.

3   Make sure that Image + Wireframe is switched on and use your mouse to position each of the images. You will need to double click each image to make it active in the Viewer, then position accordingly. Repeat this procedure for each image you wish to position. This can be done manually or, if you prefer, use the '+' symbol to set X and Y co-ordinates in the Motion Control window.

4   If you find there is a black border on any of the edges as you place the images, you need to use the Crop facility to remove this. Often a black edge is seen on the extreme outer perimeter of the video frame. This is never seen when the image is played to a television set, however, when an image is squeezed back this can become apparent. It needs to be cropped to get rid of it.

Use the slider bars to crop the image or enter a numeric value to achieve the same result.

5 Switch on the Drop Shadow in the Motion Control window. Controls for this setting are used to indicate the direction, distance, color, softness and opacity of the drop shadow.

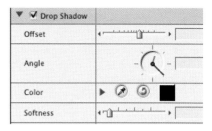

If necessary, reduce the opacity of the bottom layer to make the images on the upper layers stand out. This is achieved by using the opacity slider in the Motion window or by switching on Clip Overlays at the base of the Timeline and then moving the bar which appears within the clip to the desired level. A counter will indicate shifts in the level of opacity as you move the bar up and down.

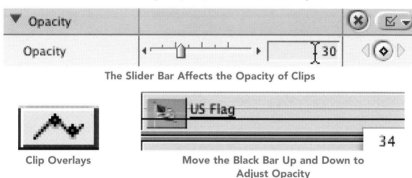

The Slider Bar Affects the Opacity of Clips

Clip Overlays   Move the Black Bar Up and Down to Adjust Opacity

It is best to plan your effects before you begin to create them. By having a clear picture of what you are trying to achieve you have a much better chance of achieving something that works. By all means experiment. It is better, however, to experiment with vision and purpose rather than stumbling around in the dark hoping something acceptable will emerge.

Keyframing Images

Keyframing allows you to set points which define a path which an effect will follow. It provides the ability to move things around, animate objects in real time and perform subtle or fast moves. An example of keyframing would be to move a box from one side of the screen to another. Effects are built through layering and adjustments to the attributes in the Motion Control tab such as size, rotation and position.

To keyframe an image you must first decide what you are trying to achieve. Let us work through the following example where we will have an image start on one side of the screen, rotate gently and increase in size until it comes to rest on the other side of the screen.

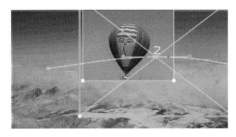

Two layers of video are involved with this example.

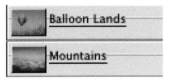

Layer 1 – Mountains
Layer 2 – Balloon

When keyframing images I change the layout of the interface. You may wish to learn this as a custom setup so that you can quickly move back and forth between the standard setup and the setup used for compositing and motion control.

The Motion tab is positioned on top of the Browser and extended to reveal its full width. Until now we have only been viewing half of the Motion tab, which is the controls section, it is now time to explore the other half of the facility which is where individual keyframes are plotted on horizontal lines. The plotted points – keyframes – represent moves which are animated over time.

EFFECTS

1. Set up the interface as illustrated.
2. Make sure you have two layers of video in the Timeline – one shot on top of another.
3. Position the Scrubber Bar on the first frame of the Sequence and turn your attention to the Canvas. The video of layer 2 will be all you see for

135

the moment. Using the pointer double click the second video track in the Timeline to make it active in the Viewer.

4   Click on the Motion tab in the Viewer.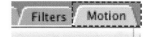

5   Use the Scale control to reduce the image in size.

6   Select Image + Wireframe mode from the pull-down menu in the Canvas. Click in the center of the image and drag the image to the left where you want the effect to begin.

Resize and Reposition the Image

7   Click with the yellow Scrubber Bar in the Timeline/Canvas and position it where you want the effect to begin.

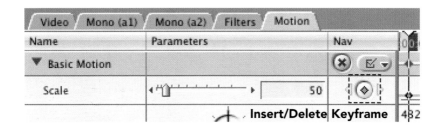

8   Press the Insert/Delete Keyframe button in the Motion Control tab of the Viewer next to the word Scale. This will set the first keyframe.

9   Use the controls in the Motion tab to rotate the image counterclockwise. You can type in a numeric value if you prefer.

10   Press the Insert/Delete Keyframe button again – to the right of the word Rotate. This will add a rotation value to the effect.

11   Press the Insert/Delete Keyframe button once more – this time next to Center. This is a positioning or location reference point.

Video	Mono (a1)	Mono (a2)	Filters	Motion		
Name	Parameters			Nav		00:00
▼ Basic Motion	Master Reset Button			⊗	☑ ▾	◀▶
Scale	◀ ⌂ ▶		50		◆	
Rotation	−⊙−		−10		◆	432 −432
Center	⊕	−238.58 ,	−91.07		◆	◇

All these points add up to one keyframe – a combination of size, rotation and positioning. The first keyframe is now marked.

12   Reposition the yellow Scrubber Bar where you want the next keyframe point to be marked. You will notice that the yellow Scrubber Bar moves in three places at once – in the Timeline, Canvas and Motion tab window.

13   Click in the center of the image and drag it to the right. A line indicating the path the image will follow will be displayed.

14   Resize the image by using the Scale controls in the Motion Control tab. The second keyframe points are automatically added as Scale, Rotation and the Center are adjusted. This is represented visually in the Motion window by the line which gradually rises to indicate the move which has been plotted.

Resize, Rotate and Reposition the Image

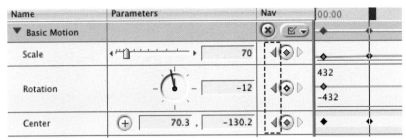

Use These Arrows to Jump between Keyframes

15 Rotate the image using the Rotation control in the Motion tab window. The second keyframe is added automatically for rotation.

You now have two keyframes plotted and this is represented visually by the dots marked on the right side of the Motion Control window. You can jump back and forth between the two keyframes by using the forward and backward arrows next to the Insert/Delete Keyframe buttons in the Motion Control window.

Playback the effect, providing the RT Extreme settings allow. Otherwise you will need to render to see the result.

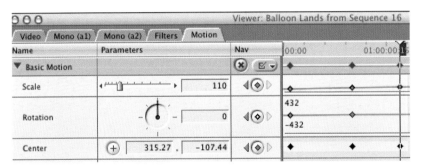

16 We are just about there. Move the Scrubber Bar forward to the final keyframe point. Adjust the Scale, Rotation and Center controls. Keyframe points for each of these will be added automatically with each adjustment.

It should be clear that a distinct set of processes is being followed each time a keyframe is added. The image needs to be sized using the controls in the Motion tab; rotated and positioned. The first keyframe must be added manually, by pressing the Insert/Delete Keyframe button, while each keyframe thereafter is added automatically each time an adjustment is made to any of the parameters.

The duration of the effect is determined by the distance or separation between the first and last keyframe. The process is logical and straightforward. It is crucial not to miss out any of the steps as this will interrupt the effect you are trying to create.

By dragging the Scrubber Bar through the effect in either the Canvas or the Timeline you can see the path the effect will follow. Playback the result through RT Extreme or render the effect, if need be. Check to see if you are happy with it. If not, repeat the process and try again.

Keyframe points can be smoothed by Control clicking on the keyframe point itself.

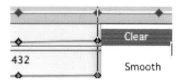

Note: Keyframes can easily be reset by clicking the Reset button at the top of the Motion Control tab. Keyframes can be deleted by pressing the Insert/Delete Keyframe button. If a keyframe already exists it will be deleted. If there is no keyframe present one will be added. Another way to delete keyframes is to select the Pen tool from the Toolbar, position it over a keyframe point and hold down the Alt/Option key (which gives you a minus symbol). By clicking on any of the keyframe points they will be deleted.

Keyframes can also be repositioned by dragging. Simply position your cursor over a keyframe point and your cursor will turn into a small cross. Drag the keyframe to a new position and release it.

You may notice that when you drag the yellow Scrubber Bar, it moves in the Timeline, the Canvas and the Motion Control area all at the same time. Each of these areas is linked, thus adjustments to one area affect the other areas.

Should you wish for the box to start outside of frame you need to reduce the size of the Canvas using the drop down % menu at the top of the Canvas.

This will resize the overall size of the frame inside the Canvas giving you the flexibility of adding keyframes outside of the television viewing area.

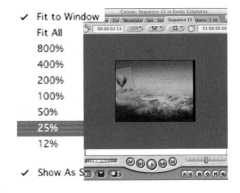

By reducing the overall display to either 12% or 25% it is possible to have animated objects fly in and out of frame from the top, bottom or either side of frame. By working with several layers of video one can animate several boxes at a time. The same can be achieved with text or imported Photoshop files.

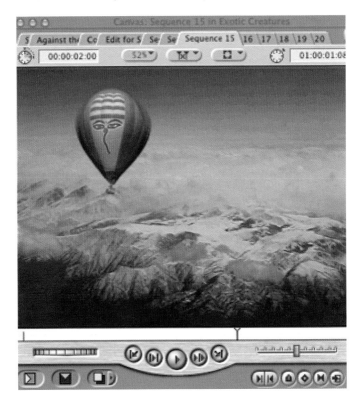

When you really know what you are doing you can make the illusion complete.

Multi-Layered Dissolves

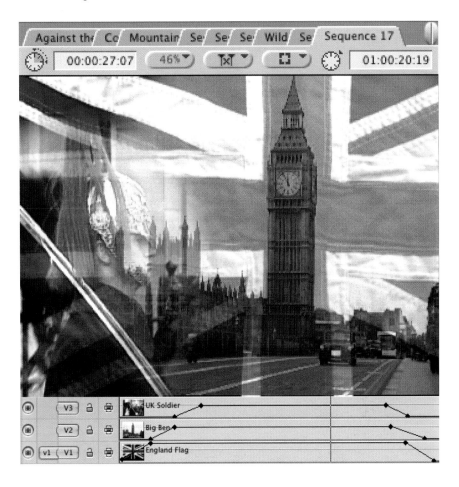

Cuts and dissolves are the bread and butter of filmmaking. Effects have always been a luxury, however, with the availability of effects which don't cost money there has been a huge increase in what clients and producers demand. It seems as if everyone wants to add extra shine to what might otherwise be a mediocre film. While overuse of effects will fail to make a bad production good – tasteful use of effects can be pleasing to the eye and if nothing else provides visual interest. The term 'eye candy' has been used to describe many of the effects offered in desktop editing systems.

A dissolve is nothing more than one shot fading into another. By placing a third layer over a dissolve it is possible to create what is known as the three-way dissolve. In the world of film production one had to plan dissolves, send the negative to the lab and hope the result was something close to what was envisaged. When using Final Cut Pro you can experiment to your heart's content. You can even create four, five and six layer dissolves and beyond. The result can end up being a mash of unwatchable imagery – so be careful.

To create a three-way dissolve is not difficult:

1. Make sure you have three empty video tracks in the Timeline.
2. Layer two clips on top of each other – the third layer will be added later.

3. Make sure Clip Overlays is switched on.

4. Adjust the opacity of the video clips on V1 and V2 to achieve the result you are after. This is achieved by moving the black line inside the clip in the Timeline up and down. Providing the yellow Scrubber Bar is positioned within the bounds of the two clips the result will be displayed in the Canvas.
5. Once you are happy with the result add the third layer of video to V3.
6. Adjust the opacity, using Clip Overlays, until you can see all three images bleeding through. You may need to make further adjustments to 'get it right'.
7. If necessary, render to check you are happy with the result.
8. Select the Pen tool from the very bottom of the Toolbar.
9. Point your cursor at the black line of the base video clip. Your cursor will turn into a pen.

10   Click on the black line and a point will be added. This will be your first keyframe.

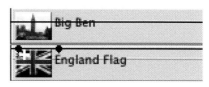

11   Add another keyframe to the left. By allowing your cursor to hover over either keyframes you will see it turns into a cross.

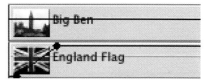

12   Drag the keyframe on the left to the base of the clip.

13   Perform the same function to the other clips, dragging keyframes to the base of the clip. What you are doing is plotting points so that the images will fade in and out at your discretion.

14   Once you feel you have the points correctly plotted, render, if you need to, and then play back the result. You may need to make further adjustments if the result is not as you wish it to be.

If you follow points 1–7 you will achieve the result of a three-way superimposition. This may be fine for your purposes. If, however, you want the images to fade in and out at predefined points you need to use the Pen tool to plot keyframes. It is possible to have a single image play, have another image dissolve over it to be followed by a third image. I'm a big fan of three-way dissolves. They can look great and produce subtle or high impact results depending on how they are treated. The interface within Final Cut Pro is particularly suited to this type of work. The hassles required in a tape suite to achieve these effects were enough to persuade many editors never to try. As for the film world – now that would have been really difficult.

Keyframing Filters

Just as Motion can be keyframed, Filters can also be keyframed. By definition keyframing means to change over time. Therefore Filters can be animated over time in a similar way to keyframing Motion Effects.

1   Add a Filter to a clip by dragging it from the Effects tab onto a clip in the Timeline. I suggest using a Gaussian Blur for the purpose of this exercise. Make sure the clip you are working with is no shorter than 10 seconds in duration.

2   Make sure the Pointer tool is selected. Double click the clip in the Timeline to which the Filter has been added. This will make the clip active in the Viewer.

3   Position the yellow Scrubber Bar on the beginning of the clip in the Timeline.

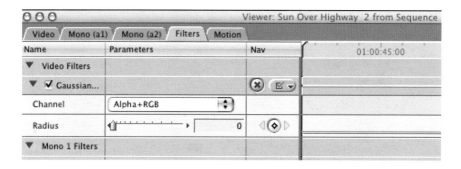

4   Select the Filters tab in the Viewer. This will reveal the controls for changing the settings of the Filter you are working with.

5   Make sure the setting for the filter effect is at 0. This will mean the Filter will have no effect on the clip at this stage.

6   Mark a keyframe by pressing the Insert/Delete Keyframe button.

7   Move the Scrubber Bar several seconds into the clip (by holding down the Shift key + the horizontal forward arrow you can skip forward one second at a time). Do not adjust the Filter settings yet.

EFFECTS

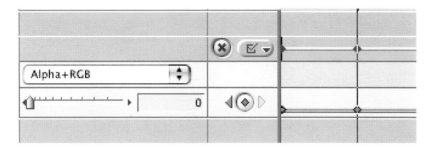

8   Mark another keyframe. You will now have two keyframes marked.
9   Move the Scrubber Bar forward several more seconds.
10 Adjust the Filter settings to bring the clip completely out of focus.

145

What you have effectively done is place a filter on a clip in the Timeline and mark three keyframes to it. The first and second keyframes are marked with the filter set to zero, thus it has no effect on the clip – this ensures when the clip plays, initially there is no visual change to its appearance. The third keyframe is added where the filter settings are adjusted to bring the clip out of focus. Thus when the clip plays, initially it is in focus until the third keyframe where it moves out of focus. This is a simple and easy way to simulate a pull focus.

By using the above method you can keyframe many of the filters found under the Effects tab. For example, you could keyframe a clip so that the appearance changes from full color to black and white – or vice versa. You could keyframe a clip using the Fisheye filter so that it appears normal and over time changes as if it is being distorted by an extreme wide angle lens. The possibilities are endless. Through keyframing one has access to powerful tools to change the look, shape and feel of clips.

Time Remapping

Time Remapping is an easy and effective technique for performing what is called 'ramping' of speed. You can 'ramp' from slow to fast, fast to slow, slow to fast to slow, and into reverse or forwards motion at will.

In essence to work with the Time Remap feature is not difficult. It involves setting keyframes and moving these keyframes to create a shift in time, within the confines of the overall length of the clip. It is important to undertstand when using this function that the overall duration of a clip remains exactly the same. When slowing down or speeding up clips using the conventional speed modifier, found under the Modify menu, the duration of a clip is made longer for slow motion or shorter for fast motion. When using Time Remap the clip stays the same in length, it is the motion within the start and end points of the clip that are affected.

Before beginning work I suggested you set up your Viewer and Browser for compositing mode – that is spread out the Viewer so that you have access to the second half where keyframes can be plotted.

1 Double click the clip in the Timeline you wish to work with. This will open the shot into the Viewer.

2 Click the Motion tab in the Viewer.

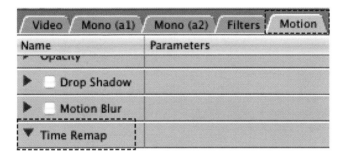

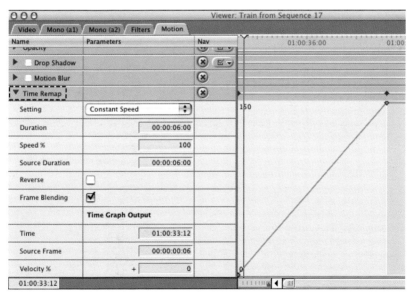

3 Click the arrow next to Time Remap and scroll down to reveal the graph on the right-hand side.

4 Click the Toggle Keyframes button at the bottom left of the Timeline. This will alter the visual appearance of the Timeline vertically. You will see increment steps below each of the clips in the Timeline.

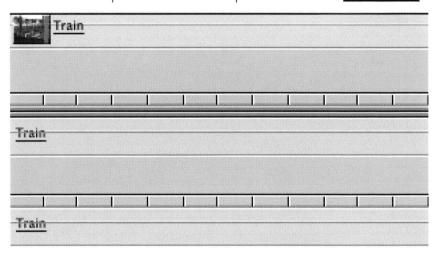

5 Choose the Time Remap tool from the Toolbar.

6 Click with the Time Remap tool at two points within the clip you wish to alter. This will plot two keyframes in the line below the clip. There will now be a total of four keyframes marked – one at the beginning, two in the middle and one at the end of the clip.

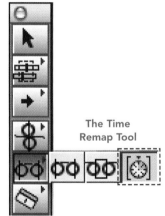

The Time Remap Tool

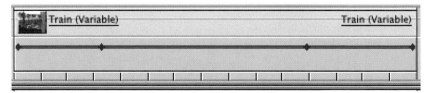

EFFECTS

Notice that the line in the graph moves from bottom to top in a continuous angle. Look to the Timeline, below the clips, and you will see that there are equally spaced increments. Both the continuous angle and the equally spaced increments indicate that the speed within the clip is constant.

Let me repeat this. If the increments are evenly spaced then speed is constant. Now for the other part of the equation – if the increments are spread out then time will be in slow motion and if the increments are close together the result is fast motion.

Slow Motion   Fast Motion

1   Select the Time Remap tool and allow it to hover within the clip, above one of the keyframe points in the center of the Timeline. The tool will turn into a + symbol.

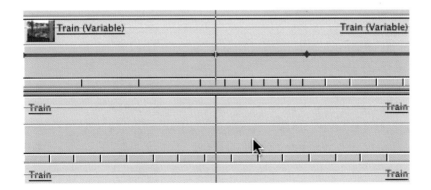

2   Click on the keyframe and move it to either side. Watch what happens to the marked increments beneath the clip (it may be advisable to increase the spread within the Timeline to get a better look at the result). You will notice the increments contract and expand as you

149

move the points. Thus, you can create an effect where a clip will begin in slow motion, speed up and then slow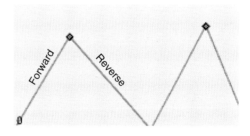
down again at the end of the clip. The plotting of keyframes is reflected in the graph in the Motion tab.

You can also move the keyframe points within the Motion window. Make sure you have selected the Time Remap tool and simply grab hold of the keyframes and move them. Keep an eye on the increments in the Timeline beneath the clip and observe whether they contract or expand. Remember, spread apart will create slow motion, whereas increments close together will create fast motion.

If you want a clip to play forward then into reverse you need to plot the keyframes so that the angle within the graphical display in the Motion tab is straight up and down.

A line climbing at an angle vertically indicates forwards motion, while a line heading downhill is for reverse. You can therefore set a clip to move forwards, backwards and forwards with each move effectively ramping time into slow or fast motion.

One final tip: if you control click on any of the keyframes in the Motion tab you will get an option to clear a keyframe or make it smooth. Select the smooth option and play back the result. You can perform Time Remapping functions at keyframe points which are smooth and fluid. This is a

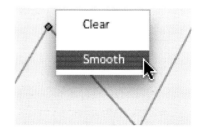

EFFECTS

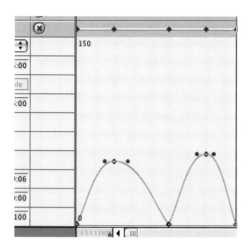

great way to make variations in motion flow in a subtle way.

Experiment with Time Remapping by using two, three, four and more keyframes. What I have described is a basic way to get this function to work. Now it is up to you to build on this.

Copy and Pasting Attributes

When you work hard at building a set of effects it can be a great timesaver to be able to copy the settings from one clip to another. For example, you may wish to run a series of clips in slow motion at 35%. Rather than setting the speed for each clip individually, it can be quicker to set this up for a single clip and to then copy the slow motion setting from one clip to all the others you wish to slow down. Details can be copied and pasted for many settings including: opacity, cropping, motion, drop shadows, filters and motion blur.

1. Control click in the Timeline on a clip from which you wish to copy the attributes. This will open a menu with many options. Select Copy.

2. Move your cursor to the clip you wish to paste the attributes to. Control click on this clip and select Paste Attributes from the menu.

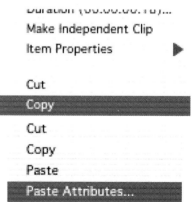

151

Paste Attributes

Attributes from Balloon Lands:
- ☑ Scale Attribute Times

Video Attributes:
- ☐ Content
- ☑ Basic Motion
- ☑ Crop
- ☑ Distort
- ☑ Opacity
- ☑ Drop Shadow
- ☐ Motion Blur
- ☐ Filters

Audio Attributes:
- ☐ Content
- ☐ Levels
- ☐ Pan
- ☐ Filters

3   Another menu will now open. Choose the options you wish to apply to the clip. Do this by checking the boxes with a tick. Obviously, the only attributes which you can apply are those which were already applied to the clip from which the attributes were copied.

Titling

As an all round editing package Final Cut Pro covers virtually every area a filmmaker will ever need to explore. No film would be complete without Titles. If nothing else a simple opening caption to identify a production is required. Final Cut Pro does much more than this including Moving Titles, Transparent Shadows and Animated Text. Some of these

EFFECTS

operations can be quite complex, however, the basics are not difficult to master.

To access Titling in Final Cut Pro go to the Effects tab and locate Video Generators.

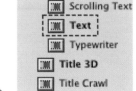

1   Click the triangle to the left of Video Generators to reveal a list of options.

2   Click the triangle next to Text – this will reveal a list of possible options to choose from.

Note: Title 3D and Title Crawl – these are additional text generators made by Boris FX. When you are ready for advanced titling, particularly credit rolls and classy looking moves, then this is where you will go.

For the moment choose the fifth option – Text.

3   Click on the Text Generator and drag it onto the V2 track in the Timeline where you want the Text to be positioned. Make sure you release it with a vertical arrow pointing downwards, thus performing an Overwrite Edit.

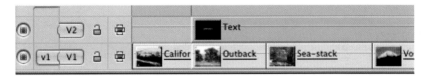

4   Double click the Text Generator on the second video track – this will load the Text into the Viewer. If your computer is 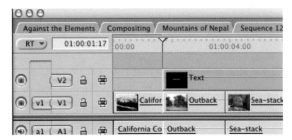 capable of RT Extreme a green line will sit above the line of text in the Timeline indicating the result will play in real time. Otherwise, you will see a red line indicating that rendering is required.

153

5 In the Viewer click once on the second tab at the top – Controls.

6 Make sure the Scrubber Bar in the Timeline is positioned on the shot which has the Text Generator positioned above it. This ensures the result will be displayed in the Canvas as you work.

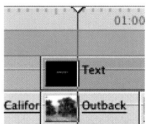

7 You should now see the words 'Sample Text' in the Canvas.

8 Return to the Viewer – make sure you have clicked the Controls tab – click in the box to highlight the words 'Sample Text'. Overtype 'Sample Text' with whatever text you wish to enter.

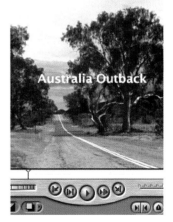

9 Click in the Timeline again and nudge the Scrubber Bar a few frames along using the arrow keys – the text will appear in the Canvas, over the shot where the Scrubber Bar is positioned.

EFFECTS

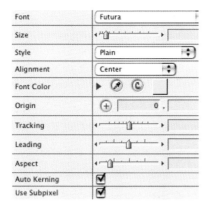 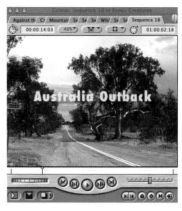

10 You can modify the characteristics of the text by adjusting the details in the Controls area. You can change the font, the size, the color and tracking by altering each of the parameters.

11 To reposition the text click on the '+' symbol in the Controls area labelled Origin. Click once on the '+' symbol and a small '+' will appear in the Canvas.

12 Clicking in the Canvas window sets the '+' wherever you click with the mouse button. Release your mouse button at the location where you want the text to be repositioned. Alternatively, you can enter X and Y co-ordinates in the Origin area in the Controls window.

Another way to move the text around is to switch on Image + Wireframe. Click with your mouse in the center of the active text window and you can then freely position the text with the mouse.

There are other Text Generators available. These include Outline Text, Crawl, Typewriter and Scrolling Text. All of these are variations on the basic Text Generator we have been working with. Experiment with these to

155

see what they do and how they work. It is also possible to layer text, as with video tracks, so that several different layers of text can be built in different styles.

The easiest way to add a drop shadow is to go into the Motion window and switch Drop Shadow on. You need to make sure that the text generator is loaded into and active in the Viewer for this to work. You can then alter the direction, color, opacity and size of the drop shadow.

The quickest way to alter the overall opacity of the text is the same as with other clips in the Timeline. Simply switch on Clip Overlays and drop the black line inside of the text generator in the Timeline to adjust the overall opacity.

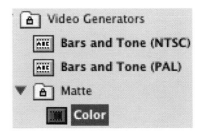

To position text over a colored background select Matte from within Generators (found under the Effects tab in the Browser). You then need to overlay the Matte Color on top of the video track and then position the Text Generator of your choice over this. The Matte Generator can be cropped in the Motion window and the transparency adjusted through either the opacity

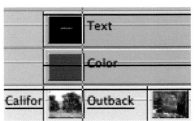

setting in the Motion window or by using Clip Overlays. Professional looking titles can then be achieved.

Switch on the Safe Title display to show what is termed 'cutoff'. This is found under the same menu as Image + Wireframe.

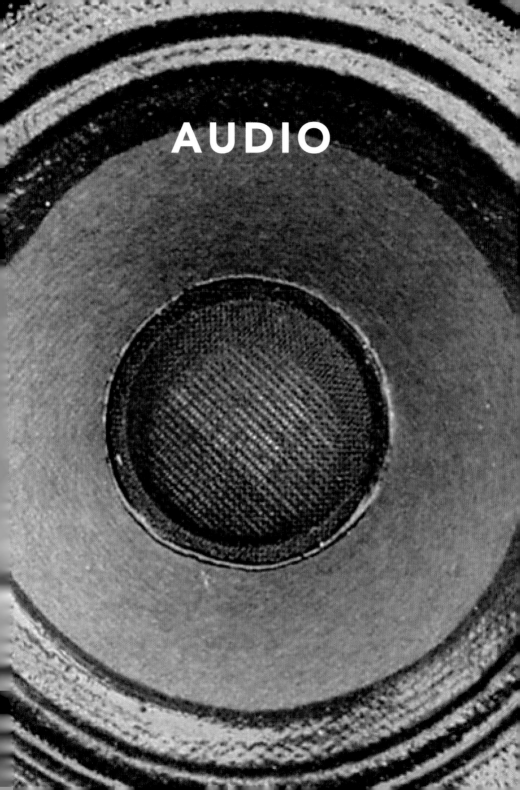

The application grows as you grow
ABBA SHAPIRO
WASHINGTON DC FINAL CUT PRO USER GROUP

It's been said before that audiences will more readily accept bad images when watching a film than they will accept bad audio. Out of focus, jerky, misframed shots, in limited doses, will not cause a viewer to switch off. Bad sound, on the other hand, will turn the film experience into a torturous ordeal. If the audience can't hear what is going on, or if the content simply hurts their ears, then it is very difficult to maintain their attention.

The only way to get good sound is to record it properly in the first place. Basic rules to achieve this include: using good microphones; getting the microphone as close to the subject's mouth as possible; listening to the sound through headphones as it is recorded and setting the audio levels correctly. The likelihood is, if it sounds good at the time of recording then it will sound good in the edit suite.

Setting Correct Audio Levels

Audio levels are crucial to get right. If you set the level too high you will blow it, literally. When digital audio peaks too loud the sound will distort, break up and be unlistenable. Often called 'pumping' this will sound far worse in a digital environment than it would have in the analog world.

The basic rule with recording sound is don't let the audio meters push into the red. This applies for recording audio on location and working with sound in the edit suite. Many experts advise that DV audio should peak no louder than $-12\,dB$. I tend to allow my audio to peak between $-12\,dB$ and $-6\,dB$ and don't experience any problems.

Working with audio in the digital domain is different to working analog. When it was all phono jacks and speaker wire one would push the audio so that it peaked high (obviously not to the point of distortion). With digital it is recommended to keep audio levels lower rather than higher. When digital distorts you really know about it. It doesn't just break up; it completely breaks up. By keeping the levels lower, rather than louder in a digital environment, you are less likely to experience problems.

Getting the Most Out of your Audio

Once your audio is recorded you are more or less stuck with it. While it is possible to improve the quality through use of filters and other means, in general the quality is determined by what is recorded on tape in the first place. However, within Final Cut Pro there are several features which enable you to put together a good sound mix. The sound mix refers to the way all of the elements blend together to create the overall soundtrack.

To produce an integrated soundtrack which is both seamless and effective one needs to be able to adjust audio levels and to program smooth fades and mix several tracks of audio together. Final Cut Pro allows the editor to adjust and mix audio levels in real time (up to eight tracks) and it does a very good job of this.

To mix audio effectively I strongly suggest that you work with what is known as Stereo Pairs. Once captured, clips can be converted into Stereo Pairs.

Converting Clips into Stereo Pairs

Each clip that you capture is made up of two tracks – a left track and a right track. When mixing these tracks, unless you are working on a complex sound mix with defined stereo separation, it is useful to marry these audio tracks together so that any adjustments to audio levels will apply to both tracks. Otherwise, when you adjust the audio levels you will need to make sure that each track is adjusted by exactly the same amount – a difficult and time-consuming process.

You can tell if your clips are Stereo Pairs by looking at the audio tracks in the Timeline. A Stereo Pair is defined by two sets of triangles facing each other. If these triangles are present then you are working with Stereo Pairs – if there are no triangles present you need to convert your tracks into Stereo Pairs.

1 Select the horizontal Arrow tool in the Toolbar and highlight the entire contents of the Timeline.

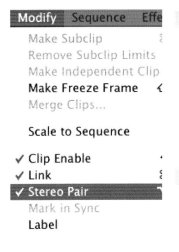

2 Choose the pull down menu at the top of the screen titled Modify. Scroll to Stereo Pair and release your mouse button. This toggles to Stereo Pair. A tick means Stereo Pair is selected. By selecting Stereo Pair you are instructing Final Cut Pro to convert whichever clips you have highlighted.

3 All your clips should now be converted to Stereo Pairs. You can confirm this by checking that two sets of triangles facing each other are present in each of the audio tracks. Providing these triangles are present then your audio has been converted into Stereo Pairs.

It is important to listen to the sound of the clips once they are converted. Then compare the sound to the clips before they were converted to Stereo Pairs. I have experienced, on some occasions, times when clips sound better as non-stereo pairs. To compare simply convert a clip to a Stereo Pair, and listen to it.

Press **Apple Z** to undo the conversion.

Press **Apple Shift Z** to redo the command.

Adjusting Audio Levels

1 Click on the Clip Overlays symbol at the bottom left of the Timeline.

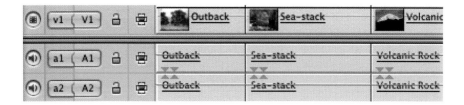

2 A pink line appears in each of the audio tracks and a black line in the video track. You will be familiar with Clip Overlays from the effects section.

3 Point your cursor at the pink lines. When you get close to these lines your cursor turns into two short horizontal lines with vertical arrows on either side.

4 Click your mouse button and you can now move the pink lines up or down. If you move the lines up the volume for the clip will increase, if you move the lines down the volume will decrease. Providing you converted the clips into stereo pairs the lines will move together as you make adjustments. Using this method it is possible to balance any differences in audio levels between clips to achieve a smooth and natural sound mix.

As you make adjustments, keep an eye on the audio levels on the VU meters within Final Cut Pro or, if you are working with a deck or a camera, on the meters on the external device.

While it is most useful to be able to adjust audio levels in this way, the requirements of most productions go far beyond being able to increase and decrease the sound levels. One needs to be able to include smooth fades and cross fades to build an audio mix where each of the elements blends together in a perfect mix of sound effects, narration and music.

Adding Sound Fades

There are two ways to add audio fades to the tracks in the Timeline. You can manually plot keyframes, using the technique known as Rubberbanding. Alternatively, the audio mixer can be used to alter the levels. Shifts in levels, using the audio mixer, can be recorded as keyframes which can then be manipulated in the Timeline.

To keyframe fades directly into the Timeline requires a similar process as adding keyframes to effects in the Motion Control window.

1. Make sure Clip Overlays is switched on and make sure all the clips in the Timeline have been converted to Stereo Pairs.

2. Select the Pen tool at the very bottom of the Toolbar.

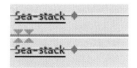

3. Point your cursor at the pink lines and the cursor will turn into the Pen tool. Choose a point on the pink lines where you wish for the sound fade to begin and click with your mouse.

A pink dot will be marked. This mark is a keyframe and represents the beginning of the fade (the keyframe will apply to both audio tracks in the Stereo Pair).

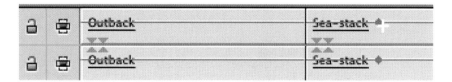

4. If you want to reposition the keyframe, point the cursor at the keyframe mark and the cursor now becomes a small cross. Click once, with the small cross on the keyframe. You can now move the keyframe by dragging.

If you wish to delete the keyframe hold down the Alt key and the
Pen tool will now have a minus symbol next to it. If you click on the
keyframe with the minus symbol the keyframe will be deleted. The
Pen tool with a minus symbol can also be selected from the
Tool Palette by extending the Pen tools and choosing the
second option.

5   Add a second keyframe further along in the clip. You should now have two keyframes marked. To create an audio fade at least two keyframes are required.

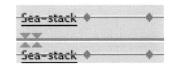

6   Hold the Pen tool over the first keyframe so that the tool

becomes a small cross. Use the little cross to drag the first
keyframe down to the base of the clip. You should now have a curved
line which starts at the bottom of the clip and rises in a curve to the
top of the second keyframe. Play back the clip and your sound will
rise from silence at the first mark to a defined volume at the
second mark.

For fine adjustments hold down the Shift key while dragging the keyframes up or down.

It is useful when working with audio to increase the size of the Timeline vertically, thus giving a better view of any adjustments you make. This is achieved by clicking on the small boxes located on the bottom left-hand side of the Timeline. Choose whichever size you feel most comfortable with.

It can also be useful to increase the overall horizontal size of the Timeline. Use the Magnifier tool to achieve this or pull on the ribbed ends of the slider bar at the base of the Timeline. This allows for fine adjustments to be made to the audio over time.

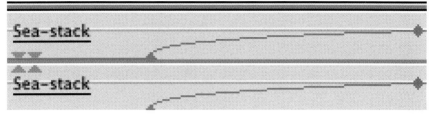

Increase the Spread of the Timeline for Fine Control

The Audio Mixer

New to Final Cut Pro with version 4 is the facility known as the Audio Mixer. This can be easily accessed from the Tools menu; alternatively select the Easy Setup titled Audio Mixing. This will set up the Final Cut Pro interface with the Audio Mixer conveniently positioned above the Timeline and to the right of the Viewer and Canvas.

The Audio Mixer allows you to visually mix your tracks, enabling you to smooth out differences in levels and to program smooth fades and shifts in volume. In essence, it performs similar functions to the Rubberbanding technique, but in a more intuitive and user friendly way. I work between Rubberbanding and the Audio Mixer in creating my mixes.

The Audio Mixer has been designed to resemble a true hardware mixer. The major difference is that it is only possible to mix a single track at a time, or two tracks if they have been converted into Stereo Pairs. With a true hardware mixer it is possible to mix several tracks, due to the fact that as humans we have ten fingers to work with and can, thus, mix several faders at a time.

This, however, is not a huge disadvantage, because it is possible to record audio keyframes, thus tracks can be mixed individually, the results recorded, and then played back while further mixing takes place.

The audio mixer is made up of several virtual faders, stacked from left to right to represent each of the tracks in the Timeline. If you have four tracks in the Timeline the audio mixer will have four faders; if you have ten audio tracks there will be ten faders for you to access in the mixer. Above each fader is a pan control, allowing you to shift the playback of audio from left to right. Tracks can be easily muted or soloed using the speaker or headphone icons. A master stereo fader is positioned to the right of the faders which controls the overall output level. A master mute button is positioned above the master fader.

A set of Audio Meters is positioned next to the Master Fader. This serves the same purpose as the Audio Meters which sit to the right of the Timeline.

To the extreme left of the Audio Mixer are radio buttons which can be used to selectively hide tracks from view. This can be useful if you are working with many tracks at a time. You may wish to hide tracks to concentrate on those you are focusing on at any given time.

The killer feature of this mixer is the Record Audio Keyframes button, located at the top right of the mixer above the Master Mute button. The Record Audio Keyframes button is used to plot points in the Timeline, in a similar way to Rubberbanding, however, the difference is the points are plotted in real time as the sound is mixed. Once these points have been plotted you can then get inside the mix, providing Keyframe Overlays is switched on, and make manual adjustments as necessary.

The amount of keyframes plotted by the Record Audio Keyframes button is adjustable. Under the User Preferences menu one can choose from the Record Audio Keyframes command. There are three choices: All, Reduced, and Peaks Only. I find the Reduced setting to be very useful. It plots just the right amount of keyframes to get inside the mix and make the necessary adjustments.

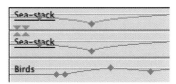

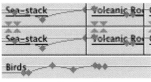

All Audio Keyframes   Reduced Audio Keyframes

Adjusting and Recording Audio Keyframes

1. Position the yellow Scrubber Bar at the beginning of the section of audio you wish to mix.

2. Click the Record Audio Keyframes button at the top of the mixer. The button will change from gray to green indicating that it is now active.

3. Play the Sequence and adjust the fader, or set of two faders which are linked as Stereo Pairs, and mix the sound in real time.

4. When you have finished your mix press the Space Bar to stop. Press the Record Audio Keyframes button to switch it off. This is important, as it is possible to inadvertently record over your mix.

If you are happy with the result carry on working, otherwise you can choose to make slight adjustments inside the Timeline using the Rubberbanding technique, or remix if necessary.

I find it useful to mix a single track at a time, or a set of Stereo Pairs, and then to repeat the procedure on another set of Stereo Pairs, or individual tracks, if necessary.

The audio level of each of the individual tracks can be raised or lowered with the Record Audio Keyframes facility switched off. Simply move the faders up or down for any adjustments. This is a quick way to bump up the level or to drop the level and provides a convenient alternative to using Clip Overlays to achieve the same result.

Adding Audio Cross Fades

Audio Cross Fades are used for creating seamless blends between Audio Transitions. All sorts or unwanted sounds can be easily eliminated.

1 Select the Effects tab located top right of the Browser.

2 Scroll down to Audio Transitions – click on the triangle to the left to reveal the contents.

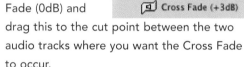

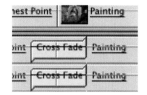

3 Click on Cross Fade (0dB) and drag this to the cut point between the two audio tracks where you want the Cross Fade to occur.

4 Playback the section with the Cross Fade and decide whether you are happy with the result. If you wish to change the duration double click on the Cross Fade which is positioned at the

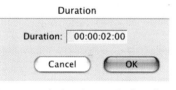

cut point between the two clips. This will open a dialog box, which will allow you to enter a new duration. Enter the duration and click OK.

The result should be a nice smooth Cross Fade where one section of audio blends into another.

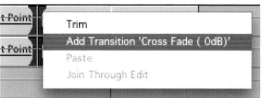

Audio Cross Fades can also be added by positioning the yellow Scrubber Bar at an edit point and choosing the Effects menu at the top of the screen. Choose one of the Audio Transitions.

169

A third way to add Audio Cross Fades is to Control click at an edit point between two clips. The option will be given to add a cross dissolve if you click in the video track or to add a cross fade if you click in either one of the audio tracks.

Adding Audio Tracks

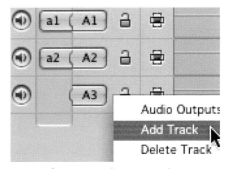

Audio tracks can be added or deleted in the same way as video tracks. Simply hold down the Control key and click in the gray area of the Timeline next to the audio symbols. A menu will open giving you the option to add or delete a track.

Remember to add two tracks for each section of stereo audio required.

Just as video tracks can be dragged directly into the Timeline the same applies to audio tracks. If you wish to add a CD track, for example, which you have already imported into the Browser, then drag the track directly from the Browser into the Timeline. If the item is dragged to an already existing audio track you will get the result of an Insert Edit if your cursor points to the top third of the track. An Overwrite Edit will occur if your cursor points to the bottom half.

Insert Edit   **Overwrite Edit**

Audio tracks can be added to the Timeline by dragging a CD track directly from the Browser to the gray area in the Timeline below the last existing audio track.

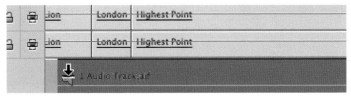

The CD track will then be added and two new tracks created to accommodate it.

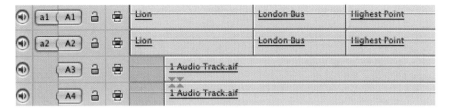

This is the quickest way to create new tracks and at the same time get a new piece of audio into the Timeline. If you wish you can then delete the audio and these tracks will remain free to be used however you wish.

Mixdown Audio

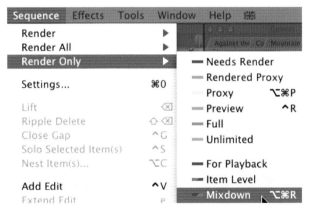

If you work with more audio tracks than your computer can comfortably play then you may choose to invoke the Mixdown Audio command.

1. Select the Sequence menu and scroll to Render Only.
2. Move right and down until you get to Mixdown.

The Mixdown command can also be accessed by pressing Apple + Alt/Option + R.

This will mix all your audio tracks into a single file. Everything will appear exactly the same in the Timeline, however, Final Cut Pro will reference to a single file. This is useful if you are working with a complex audio mix which involves more tracks than your computer can mix in real time.

171

The downside to using the Mixdown Audio command is that each time you change the edits in the Timeline you will need to do another mixdown. Unless you are working with many tracks of audio, or use a lot of audio filters, keyframes or effects, you may never need to use this facility.

> *I can't tell you how amazed I was when I did my first output from Final Cut Pro in the living room of my apartment. And I was dancing all around the room just as happy as can be.*
> KEVIN MONAHAN
> SF CUTTERS

Once your production is complete it is time to output your movie.

The simplest form of output is to roll record on your camera or deck and then play the movie direct from the Timeline and straight to tape. This is something you should do regularly throughout the editing process to ensure you have a backup of your film should a technical catastrophe or any other sort of problem render your project useless.

I always back up my material to digital tape in the form of several different versions: with and without graphics and effects and with separate audio tracks, both mixed and unmixed. One pass may include voiceover, another music, and another sound effects. By doing backups in this way all of the raw elements of the film are preserved so that in the event of a corrupt file, human error, or power failure, you will be able to rebuild your film from the elements recorded to tape. It takes discipline to back up your work regularly, however, unless you have tremendous amounts of hard drive space, so that you can afford a true backup of all of the project files in the form of data, then backing up to digital tape is the best option.

Print to Video

Print to Video is a function Apple built into Final Cut Pro to give a professional look to a finished film. Here the project can be named, color bars can be inserted at the head and black can be added at the end of the production. It is also possible to loop the film so that several copies can be recorded onto a single digital tape, or one can choose to record sections of the movie by defining 'in' and 'out' points and only printing these sections to tape.

Print to Video also serves another purpose. If any material is unrendered then the computer will render this material prior to invoking the Print to Video command and if your audio mix is complex then a mixdown will take place.

This ensures all the components of your production should play without problem. Sometimes dropped frames may be encountered during normal playback and the problem will be cured when the Print to Video instruction is given.

Selecting Print to Video is often the final stage of the editing process. Think of it as getting the release print off to the lab once the hard work in the cutting room has been done.

1 Select the File menu at the top left of Final Cut Pro and scroll down to Print to Video.

2 Release the mouse button and a box will appear giving you many options to choose from. Check those boxes which apply to your specific needs.

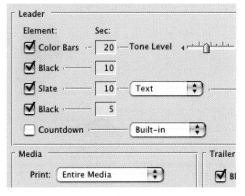

Add color bars if you wish and enter a duration for the amount of bars you wish to record. Instruct Final Cut Pro if you want black to be recorded after the bars. Enter a title for your production in the text column and select Print Entire Media or Print In to Out depending upon your requirements. If you choose the Print In to Out option you need to mark 'in' and 'out' points in the Timeline.

3 Once you are satisfied that you have correctly selected the options you require click the OK box. The computer will pause as all the elements are gathered together.

Now sit back and enjoy your movie. It is worth keeping a close eye on the output. Print to Video is the last stage in the editing process (unless you are using it to label sections of a work in progress) and therefore it is wise to make sure everything is exactly right as the signal is recorded onto tape. Once you have at least one master safely dubbed you can relax knowing that the vital information has made it from the original camera masters, onto your computer's hard drives, through the editing process, and finally back onto digital tape. Don't forget to do a second and maybe even a third backup. This may just help you to sleep better at night.

Other Forms of Distribution

Final Cut Pro is designed as a video editor, however, just because it is designed to work with video does not mean that your final output must end up on videotape. There are other formats of distribution in this modern world that need to be considered; for example: DVD, CD-ROM and Internet delivery.

For most of these purposes, unless you are working uncompressed, one would acquire and edit the material on DV or DVcam tape and output the master edit to digital tape before compressing the final product to another format. By having your file securely backed up on tape you can export the file from Final Cut Pro to the format of your choice, knowing the master edit is securely recorded to tape.

QuickTime is a cross-platform (Mac and PC) technology designed and developed by Apple. This is often the format of choice for both Internet delivery and CD-ROM production. QuickTime is powerful and achieves results. It is particularly suited to producing encoded files on a Mac for the simple reason that QuickTime is part of the Mac operating system. It had been described as the 'the thing' that makes video on the Mac possible. However, before choosing the exact form of compression, you need information as to how and where the product will be used.

When outputting to DVD, MPEG-2 is the compression format which must be used.

To output compressed files using Final Cut Pro you can choose to export using Compressor, which runs as a separate application to Final Cut Pro, or you can export Using QuickTime Conversion.

Export Using Compressor

The main function of Compressor is to encode files to MPEG-2 for DVD authoring. This is obvious from the list of presets available in this application. Compressor can also be used to compress files to QuickTime, AIFF, or MPEG-4.

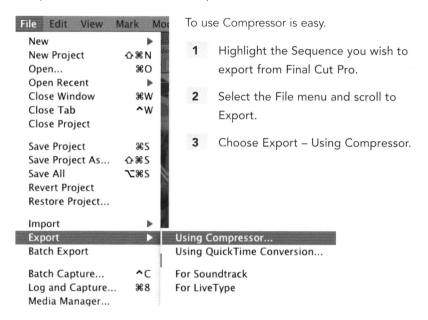

To use Compressor is easy.

1. Highlight the Sequence you wish to export from Final Cut Pro.
2. Select the File menu and scroll to Export.
3. Choose Export – Using Compressor.

The Compressor window will open in front of you with the name of the Sequence you have chosen named in the Source Media column. You now need to choose a preset.

4 Click the opposite facing arrows in the Preset column and choose one of the available options. It is best to choose the highest quality setting possible. There are options for Fast Encode, however, unless you are restricted for time, or space, choose one of the high quality settings.

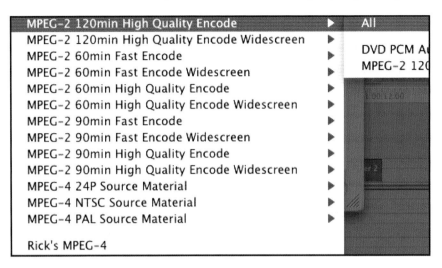

Assuming you are encoding to MPEG-2 for DVD authoring, then two files will appear in the Preset column. This is because audio and video are encoded as separate files which will then be used inside the DVD Authoring application when you get to that stage.

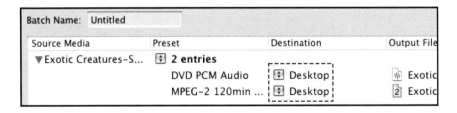

5 Click the opposite facing arrows in the Destination column and choose Desktop for each of the files listed. This will save the compressed files to the desktop where you can then access them and file them away. If you prefer, select Other and navigate to a folder or location where you wish the files to be stored.

6 If you have other files to compress, return to Final Cut Pro, highlight a Sequence and export for Compressor. Repeat all the steps you have just gone through, including choosing a preset and selecting a destination.

7 Submit the batch for encoding.

The files will now be processed. If you wish to stop the Batch for any reason, highlight the Batch Name in the Batch Monitor window and press Stop. You can then choose to delete the Batch or Resume the encoding process.

If you wish to encode audio files to AIFF, or video and audio to MPEG-4 for Internet or CD-ROM production, then you follow exactly the same instructions, however, substitute the MPEG-2 preset with one of the AIFF or MPEG-4 presets.

At any time prior to submitting the Batch for encoding, you can choose to Preview any of the files in the Batch. Simply highlight the file within the Batch window, by clicking in the Preset column, and press the Preview icon. This will open a window which will provide a vertical split which can be shifted from side to side, allowing for a direct comparison between the original and compressed file. Using this facility will give you some idea of what to expect when the file is encoded.

Within the Batch window you can easily add or delete files by pressing the + and − symbols at the bottom left of the window. It is thus possible to line up as many files as you wish into Compressor, so that you can then submit the Batch and let your machine do the work while you take a break or leave the office, knowing that the encoding is being completed while you deal with other matters.

Export Using QuickTime Conversion

When compressing video for the web, CD-ROM or DVD, decisions must be made about what level of compression will be used. When compressing large amounts of data there are always compromises which must be made. For example, when preparing material for Internet delivery the goal is to reduce the overall file size to a minimum while maintaining the best possible quality. Most consumers use a 56k modem and therefore sending gigantic video files down a telephone line is simply not an option. Even still images are compressed heavily for web use and video is made up of thousands and millions of stills.

The purpose of this section is not to give detailed instructions of how to achieve these forms of output, what levels of compression to use, and exactly how to go about this – but rather to give you an overview of the areas you need to explore. There are entire books written about compression and the best ways to achieve results. What I will provide here is an introduction to the compression process using QuickTime as a means for facilitating the conversion process, and how it can be achieved using the tools provided with Final Cut Pro.

1 Highlight a Sequence or clip you wish to export in the Browser of Final Cut Pro.

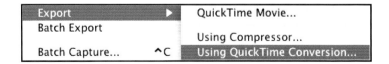

2 Choose the File menu, scroll to Export. Choose Using QuickTime Conversion.

3. Look to the bottom of the window that opens. Next to the word Format is a bar labelled QuickTime Movie. Click this bar to reveal a selection of options. You need to choose the setting appropriate to the purpose you are working towards.

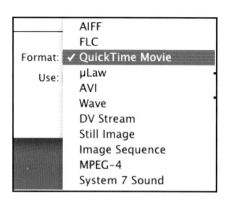

Some of these options were available using Compressor, however, there are other options which are not. For example, you can export as an AVI file – used for creating files compatible with Windows Media Player or you could choose to export a single frame as a still image.

Be aware that under each of the possible compression choices are further options. For example, if you choose QuickTime Movie and click the Options setting to the right, then you can choose the quality of the video, which format you wish to compress the video to, and the audio sample rate.

You need to experiment. Look through the options, make a choice, export a movie and see the results. You can set different levels of compression, different samples rates for sound, decide whether the movie is to be prepared for Internet streaming or not. There is no easy option. You really need to know what you are trying to achieve before you start.

And, as you can see to the right, there are other possibilities to consider beyond simply choosing QuickTime.

Audio, for example, can be exported as an AIFF file. When exporting as an AIFF file consider: is the sound mono or stereo? what is the sample rate?

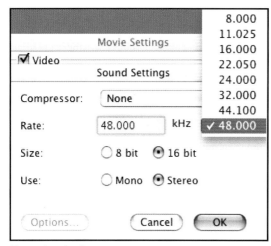

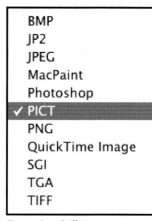

Exporting Audio

Exporting Still Images

Images can be exported as Still Images. Choose PNG, TIFF, Pict or Photoshop to name just a few of the options.

If all this sounds confusing it is because we are dealing with a confusing subject – video and audio compression is constantly evolving and therefore changing. However, having access to the tools gives you access to the knowledge you need to work within this area. The beauty of working on the computer is you always get a second shot at it, and a third, fourth and hundred and thirty-first shot if you need it. A little bit of knowledge will take you a long way and your knowledge will accumulate as you go. Remember – knowledge is power! The power of distribution is at your fingertips through the process of compression. Press the buttons, wait for the encoding, and see the result.

> *The digital video revolution has just begun.*
> *Final Cut Pro is making us all leaders of that revolution.*
> DAN BERUBE
> BOSTON FINAL CUT PRO USER GROUP

Titles on early films were primitive when compared with those of today's productions. Titles were originally made up on cards or rolls of paper which were then filmed directly in front of the camera. More sophisticated titles were created using innovative methods such as painting onto glass and filming live action in the background. Later, the optical printer enabled the level of sophistication to reach new limits.

As television evolved Letraset became widely used with titles being created letter by letter. Blades, glue and card were the tools of the graphic room which created artwork which was placed directly in front of a camera. The result was then keyed over live or taped action.

The 1980s and 1990s saw a wave of dedicated computers, known as character generators, which were used specifically for the creation of still and moving titles. Character generators are still used in live and post-production environments today.

With the development of non-linear editing systems titling has become a standard part of the functionality of the editing application. With Final Cut Pro this is catered for with the in-built text generators and the two additional plugins supplied by Boris FX.

Enter LiveType. Now we can all do network television graphics.

LiveType is a separate application to Final Cut Pro. As such, video must be exported from Final Cut Pro to provide a reference to position text over.

The capabilities of LiveType go beyond providing a facility capable of producing text. Moving backgrounds can be created, over which text can be positioned, and text can be manipulated in a variety of ways. Letters can be individually animated and layered, and a variety of effects can be applied to still or motion graphics.

The very first thing I do whenever I open LiveType is to set the Project Properties. This is found under the Edit menu.

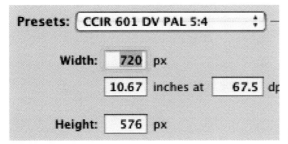

You are now aware that LiveType is a completely separate application from Final Cut Pro. As such, if you want to work with moving video, or replace the video in any of the templates, first you have to export it from Final Cut Pro and then bring it into LiveType.

1. Highlight a Sequence in Final Cut Pro. If you wish to export the entire Sequence clear the 'in' and 'out' points. If you wish to export a section – mark the 'in' and 'out' points and this will define the section to be exported.

2. Choose the File menu, scroll to Export and choose For LiveType.

3. Name it and save it to the desktop.

LiveType Templates

To introduce you to the capabilities of LiveType it is best to have a look at the templates that are included. All up there are 60 templates to choose from and these are broken into various categories.

185

1. Open the LiveType application.
2. Select the File menu and select Open Template.
3. Click on any of the categories to reveal the pre-built templates inside LiveType.

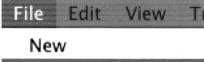

Click through the templates. Watch the various possibilities flow before your eyes. Choose a template by clicking OK.

Some of the 60 Templates Available in LiveType

You may wish to search through each of the four categories. You can choose between full frame graphics with moving text and video, lower third name graphics, and a selection of promo and title backgrounds. Select a template!

Replacing Text in the Templates

Look in the Timeline. It is easy to identify which tracks have been used for the construction of the layers of text within the template.

1 Double click the track in the Timeline.

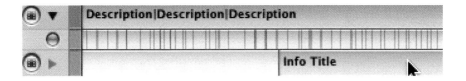

2 Overtype the text in the box top left of the Inspector window. As you type the result will be obvious in the display in the Canvas.

The templates are customizable which means you can replace any of the details with your own information. For example, you can change the words, images and other attributes such as style and color.

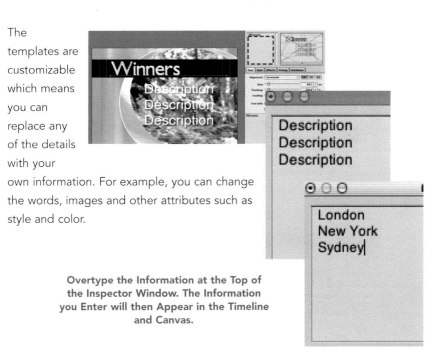

Overtype the Information at the Top of the Inspector Window. The Information you Enter will then Appear in the Timeline and Canvas.

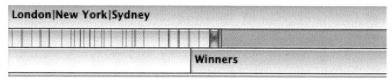

The Information You Type in the Inspector is Reflected in the Timeline.

Replacing Video in the Templates

Not all the templates use moving video. But for those that do, you will want to know how to change it.

1 Go to the track furthest down the hierarchy of the Timeline.

2 Highlight it.

3 Delete it.

4 Select the File menu.

5 Scroll to Place Background Movie.

6 Select the file on the desktop which you exported from Final Cut Pro earlier on.

7 This will import the video and insert it into the Timeline on the track below the line.

Resize the Canvas if necessary by clicking the percentage symbol located bottom left

8 Press the play button at the bottom of the Canvas or click in the Timeline and press the Space Bar. This will perform a render of the effect into memory. Then the effect can be played back in real time.

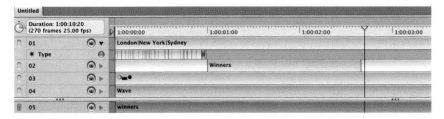

Working Inside LiveType

The templates are just the beginning of LiveType. When you dive into using the manual controls, you will find the power to manipulate text in combination with Backgrounds and Textures.

It is now time to move away from using the templates and onto using LiveType as a compositing tool.

1 Create a new project. Select File – New.

2 Set the Project Properties, found under the Edit menu. Choose the format you are working with.

3 Click in the text area of the Inspector and type some text. As you type the letters will appear in the Canvas.

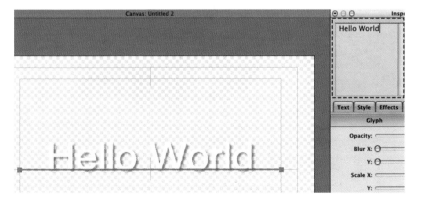

189

4   Grab the blue bar with your mouse and you will find that the text can be freely repositioned.

5   Move the points at either end and the text can be slanted.

There are five tabs in the Property Inspector. This is where you control the settings that relate the text.

Text – you can set the text size, tracking, leading and alignment.

Style – this is where Shadows are controlled. Shadows can be blurred, scaled, offset and changed in color. One can add glows, outlines and extrude the shadow information.

Effects – displays a list of effects which have been added to the text in the Canvas.

Timing – parameters for timed motion effects can be adjusted here.

Attributes – here text can be blurred, scaled, rotated and changed in color.

One of the powerful features about LiveType is that each letter is effectively positioned on its own layer. Letters can be adjusted individually or together in a group.

Media Browser

Five tabs line up at the top of the Media Browser giving you access to the media which is bundled with LiveType.

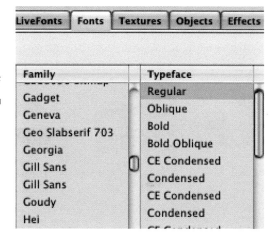

Livefonts – these are pre-built animated fonts. Click on any of these and the preview window of the Media Browser will demonstrate the animation you can expect.

Fonts – the building blocks of typography – the fonts are provided in Family and Typeface descriptions.

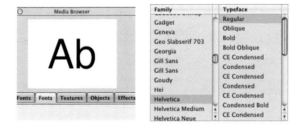

Textures – solid backgrounds to place behind your words. Click on any of the textures and they will appear in the Media Browser window.

Objects (see overleaf) – these are moving elements which blend together well with other graphics. Again, click and see the result.

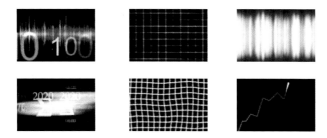

Effects (see below) – pre-built effects which you can readily apply to your text to give it movement. Click on each of the settings to see the result.

Any of the choices under the Media Browser tab can be applied to the text in the Canvas, or placed behind it as in the case of the Textures and Objects.

By selecting and modifying the text in the Inspector, and then using the tools in the Media Browser, it is possible to build titles with those characteristics demonstrated in the Templates. The main difference is that you build the graphics as you want them to be.

As you apply text, add backgrounds, import video and make use of text and objects – all of these elements are layered in the Timeline to represent your composited elements.

You need to work to interpret the relationship between the elements in the Timeline and the Canvas.

Just remember, whatever you create in LiveType will be made up of a combination of Livefonts, Fonts, Textures, Objects and Effects, with the Inspector serving as the place where attributes, size, and timing (plus a whole lot more) are adjusted.

Creating Results Manually in LiveType

We will now explore three possible graphic compositions which can be created in LiveType.

(i) Text over a textured background.

(ii) Text over live video.

(iii) Animating text over a background or moving video.

To create text over a background is quick and easy to achieve.

1. Type the text in the Text window of the Inspector.

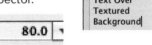

2. Adjust the size of the text.

3. Select the Attributes tab and set the color of the text.

4. Position the text by moving the blue bar. It may be helpful to reduce the overall size of the Canvas to get an overview of the workspace.

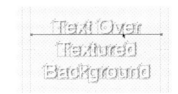

5   Click the Textures tab in the Media Browser. Choose a texture you like.

6   Press Apply to New Track.

The text you have created is now positioned against the textured background.

To create text over moving video is just as simple; first you must export the section of video from Final Cut Pro so that LiveType can recognize it. This process has already been described at the beginning of this chapter.

1   Export a sequence of video for LiveType from Final Cut Pro and save this to the desktop.

2   Create a new document in LiveType.

3   Type the text you wish to work with in the Text window in the Inspector.

4   Select File and scroll to Place Background Movie. Navigate to the video you wish to import on the desktop.

5   Navigate to the video you exported earlier to the desktop; highlight it, and click the Open command. The video will now appear in the Canvas as a layer below the text you typed earlier.

6   You can now manipulate the size, color, positioning and other aspects of the text using the controls found under the five tabs in the Inspector window.

To add a moving effect to the text you have created, regardless of whether it is over a textured background or video which you have placed in the background, then simply access the effect from the Effects tab and apply it.

LIVETYPE

1 Make sure the text is selected. Click the blue bar in the Canvas to select it.

2 In the Media Browser click the Effects tab. Click on the various effects to see what they do.

3 Choose an effect and press Apply. This will apply the effect to the text in the Canvas. If you look to the Canvas you can see three tracks. The base layer, below the line, is the video background. On the top layer is your typed text. In the middle is the effect.

4 Press Play in the Canvas, or click in the Timeline and press the Space Bar. This will cache the effect into ram and enable you to play it back in real time to check the result.

5 The duration of the effect can be changed by going into the Timing tab in the Inspector and moving the slider bar next to Sequence. Alternatively, drag the end of the effect in the Timeline to increase or decrease the overall duration.

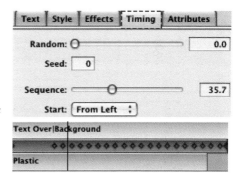

You can then apply the characteristics of any of the Livefonts in the Media Browser.

1 Select the text by clicking the blue bar in the Canvas – alternatively, highlight the text in the Inspector window.

195

2 Click on the Livefonts tab.

3 Click through the options and choose an effect you want to work with.

4 Press Apply. The text in the Canvas will now show the characteristics of the Livefont you have selected. Press the Play button in the Canvas, or click and press the Space Bar play to send the frames to Ram for playback.

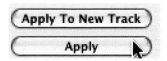

Output from LiveType

Once you are happy with your creation you need to select File – Render Movie. This will build each of the frames in the LiveType Sequence.

1 Now choose Export Movie from the File menu. Name it and save it to the desktop.

2 Once the file is saved go back into Final Cut Pro, import the file you have just exported from LiveType, and from then on work with your creation inside Final Cut Pro.

Note: When you choose the Render Movie command a dialog box will appear giving you the option to either Render Only Between In/Out Points or to Render Background. Check Render Background if you wish for either the background video or a textured background to appear as part of the file which you will then import into Final Cut Pro.

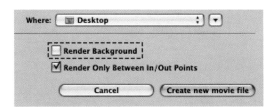

Once you have positioned the file in Final Cut Pro you will need to render the result. If need be, manipulate it further inside Final Cut Pro with regards to positioning, cropping, size, or any of the facilities available to you inside the Motion tab, Timeline or through further use of transitions, filters and generators.

LiveType is an application you must teach yourself. The possibilities go on forever. It has a rich, powerful and customizable workflow which can be used, in combination with Final Cut Pro, to create multi-layered graphic sequences which only a few years ago would have required dedicated hardware specific to text creation. In my opinion, LiveType is the most sophisticated and capable character generator I have ever worked with!

> *Everything you need is at your fingertips.*
> RICK YOUNG
> UK FINAL CUT USER GROUP

When the very first films were made there was no way of recording a soundtrack. In theatres throughout the world musicians would play live music to accompany the film. As technology evolved various methods were employed to enable recorded material to be played in sync with motion pictures. However, it was still necessary to hire musicians to compose and play the musical score which would then be recorded and integrated into the sound mix of the film. Later, stock music became available to filmmakers who couldn't afford the expense of an orchestra or band of session musicians. Stock music would be purchased on vinyl, and later CD, and each track would have to be logged and accounted for as the films were distributed or broadcast on television. Royalties would then be paid to the composers who created the stock music. In recent years, what has become known as royalty-free music has become available. Royalty-free music requires a one-off purchase which would then give the purchaser unrestricted use of the music. This cut down on the red-tape of the copyright game, enabling filmmakers to buy a compact disc and use the music in as many productions as one wished.

Apple have taken the idea of royalty-free music one step further in creating Soundtrack. This innovative program provides a series of musical loops and pre-recorded pieces which can be ordered, layered and mixed in any way the operator wishes. The tempo, key and form of the music can be altered, worked and reworked in a myriad of ways to create any range of musical scores. There are no musicians to pay, no forms to fill out, and no fees to pay. It is all bundled with Final Cut Pro 4 as a stand-alone application.

Now before we get into using Soundtrack just a word on the system requirements. You need at least a G4 500 MHz machine with 1 MB of cache to run Soundtrack. So if your machine is below this benchmark then you are out of luck.

IMPORTANT: before you begin

Before you can access the musical loops, which are the elements you will use to build your musical score, you need first to search the database and index the directories for the loops and other audio files that you will work with. This is a one-off process which, when done, will be remembered by Soundtrack each time you open the application.

To search the database and index the directories:

1   Open up Soundtrack.

2   Click Search – the third tab to the right on the top left of the interface.

3   Click the Setup button.

4   Highlight the hard drive listed in front of you. Otherwise click the + button and navigate to the drive where the media was installed during installation.

5   Press the Index Now button. Once the indexing is finished press Done.

Now, whenever working with Soundtrack, or LiveType, you are dealing with programs that are separate to Final Cut Pro. It would be nice if everything integrated into a single program but that is not the way it works. Soundtrack functions as a completely separate application to Final Cut Pro.

Exporting for Soundtrack from Final Cut Pro

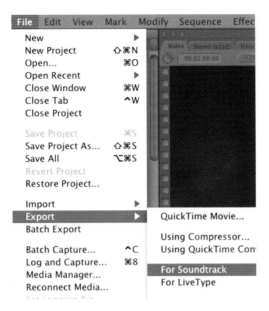

To get a Sequence file out of Final Cut Pro, so you can work with it in Soundtrack, you need to select Export for Soundtrack. The process is the same as Exporting for LiveType, except you choose the Soundtrack option.

1. Select a sequence in the Browser of Final Cut Pro. If you want the entire movie to work with clear the 'in' and 'out' points in the Sequence. If you want a section of the Sequence then mark the 'in' and 'out' point for the section of the Timeline you wish to Export.
2. Highlight the Sequence and choose Export – For Soundtrack.
3. Save the file to the desktop.
4. Open the Soundtrack application.

SOUNDTRACK

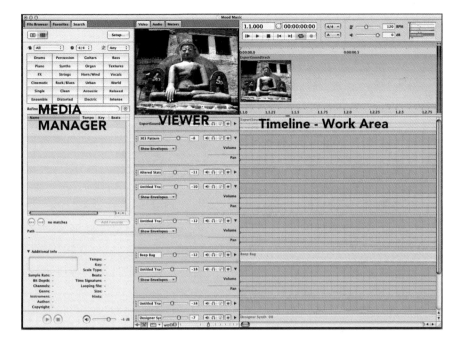

The interface is made up of three main windows:

Media Manager – this is where you access all of the components from which you will build your musical score.

Viewer – your movie will play in the Viewer as you build the music to match the picture.

Timeline – as in Final Cut Pro, the Timeline is the area where the building blocks are edited. You can layer as many tracks as you need to work with in the Timeline and the controls such as Volume, Pan, Transpose and Tempo can be adjusted and keyframed over time.

In many ways Soundtrack is similar to Final Cut Pro. Apple have deliberately shaped the program so that there is immediately some degree of familiarity.

The layout of Soundtrack can be easily set up by selecting the View menu and selecting Layout.

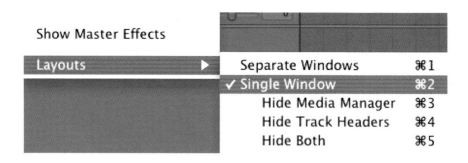

For the moment I suggest working with the default layout which is Single Window.

Icons at the base of the Timeline provide easy access to useful functions:

Snapping – toggles on and off.

Four boxes allow you to alter the size of the tracks you are working with.

A convenient slider bar allows you to increase or reduce the size of the tracks.

Providing you have searched the database and indexed the directories you are ready to begin work on creating the musical score for your project.

The Soundtrack Workflow

There are four steps to the Soundtrack workflow.

Export file for Soundtrack from Final Cut Pro.

Import this file into the Viewer of Soundtrack.

Build your musical mix.

Export the tracks so they can be used in Final Cut Pro.

SOUNDTRACK

Importing a Movie into Soundtrack

This is an important and simple stage as it will allow you to preview your edited movie in the Viewer while you work.

The easiest way to get the file into the Viewer is to drag it straight from the desktop and into the Viewer.

Locate the file you exported earlier from Final Cut Pro and drop it onto the Viewer inside Soundtrack.

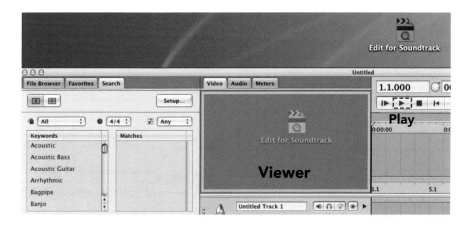

Press the Play button to hear your movie. You will now see your film play in the center window and hear the sound which you exported from Final

205

Cut Pro. A waveform of the audio will be displayed in the track area of the Timeline.

Building your Mix

Basically Soundtrack is a library which provides easy access to many loops, pads, chords and sound effects. With everything from electric piano, guitars, drums, bagpipes and bongos, Soundtrack provides a full range of choices.

1. Go to the File Browser which is the library of all your pre-recorded components.

2. Click the third tab at the top of the Browser – Search.

3. Toggle between two possible displays of the list of components you have to work with. You can choose to view the components As Columns or As Buttons.

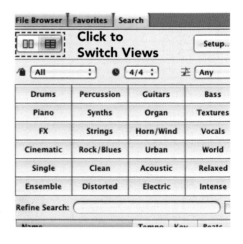

4. OK – now it's time to start listening to the music. Regardless of which mode you display all the components in, you need to click on a description. This will provide you with a list in the center of the

SOUNDTRACK

Browser. Each of the loops is named for reference purposes. Click on a loop and listen to the sound. To stop a loop from playing click the Stop button at the bottom of the Browser. Click through the various loops and listen to as many sections as you wish. You will get a feel for what elements exist in the library.

Note: if you cannot hear anything check that the volume is turned up on your Mac. Ideally, you should have a set of external speakers plugged into the output at the back of the Mac. If you have no external speakers you will have to listen through the internal speaker inside the Mac. Alternatively, use a set of headphones.

5   This is the fun part. When you find a piece of music you like, pick it up and drop it into the Timeline. Repeat the process as many times as you wish – as you move inside the Timeline you will see your movie play in the center of the interface.

Drag a Loop from the File Browser to the Timeline Work Area

207

6   Layer sound elements on several tracks. New tracks are easily created by dragging files directly from the Browser into the Timeline.

7   Drag the ends of the loops to make a continuous track for whatever duration you require.

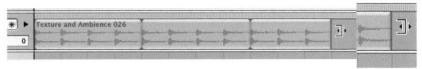

Position your Cursor on the End of a Loop and Drag the End

8   If you wish to increase or decrease the size of the tracks choose any of the boxes at the base of the Timeline. You can use the slider bar to increase or decrease the overall spread of the Timeline.

9   Search through the library and build a musical structure. You can combine drums, keyboards and guitar. It may be formulaic but it works. Often what is needed in a film is music which is subtle, hidden in the background. You may barely want the viewer to notice it, or make it jarring at just the right moment for effect.

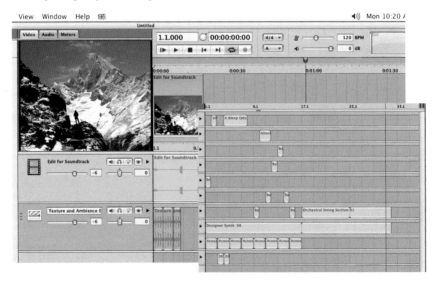

10 The volume of the individual tracks can be easily raised or lowered. There are two methods: (1) move the volume slider from side to side (2) click on the arrow to extend the track and you have something that looks remarkably similar to Clip Overlays. Volume and Pan controls are accessible through clicking with your mouse and sliding the line up for an increase in volume or down to lower the sound.

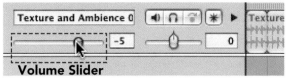
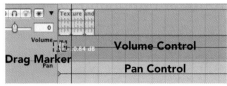
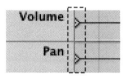

11 The two controls beneath the tracks can be keyframed. Fades are easily programmed by double clicking with your mouse and then dragging the point that is added.

12 Look to the top of the display. Notice the BPM button. BPM stands for Beats Per Minute. Drag the slider backwards and forwards and listen to the result at the various positions. **Note**: the Beats Per Minute can be varied for loops already positioned in the Timeline, or loops can be adjusted before they are dropped into the Timeline. If you forget what the original settings were then simply refer back to the name of the original loop in the Browser. Next to the name is the Tempo, Key and Beats per minute in the description column.

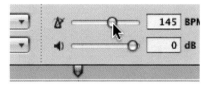

13 You can also alter the volume, tempo or transpose a track by clicking the Master Envelope button at the base of the Timeline. By pressing this button the Master Envelope will toggle on and off.

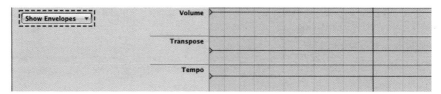

Show Envelopes Reveals Volume, Transpose and Tempo Controls.

Exporting the Mix

Once you have a result that you are happy with you then need to export the mix.

This can be done in one of two ways: (1) Export Mix (2) Export Tracks.

If you choose to export the mix a single file will be created – I suggest you export the file to the desktop. The file can then be brought into Final Cut Pro and edited into the Timeline like any other audio file.

If you chose Export Tracks, you need to first create a folder to store the tracks. Create a folder on the desktop and export the tracks to this location.

The way the Export Tracks command works is clever. The full duration of each of the tracks is exported, including gaps of silence, which makes it easy to line up the tracks individually to work with inside Final Cut Pro.

Once exported from Soundtrack import the file or files into Final Cut Pro and line up these elements with your original movie.

That's the Soundtrack workflow. This program will keep you busy for a long time – there's plenty to explore, plenty to learn.

Soundtrack is a complete application. It is no toy or gimmick. It is, without doubt, a professional program suitable for use in the real world of film and video production.

Epilog

Apple have always been individual in the way they do things. The way their machines operate, the way they look, and even the chips inside are unique. No matter how hard the competition tries, Apple has remained a formidable force in the creative industries of the world.

The reason why Apple continues to be a success is the result of an extremely loyal customer base who refuse to abandon their platform of choice. People do not remain loyal to Apple out of some kind of blind devotion – they remain loyal because Apple machines and software serve them better than their competitors.

People like using Apple computers.

People love using Apple computers.

When it comes to Final Cut Pro, the same applies.

In a few short years Apple have created an editing system which shines bright across the murky waters of the post-production scene.

Final Cut Pro is reliable and stable. It works very well. And it works with every known format on the planet: video and film.

It is the success it is for very clear reasons:

It's easy to use.

It's powerful.

It's affordable.

A friend said to me years ago: 'You know, I don't know why Apple computer's still exist.'

They exist because the people want them to exist.

That's it.

Finally, Cinema Tools. This application serves two purposes: (1) to enable you to output film cut lists (2) to allow you to edit video at 24 frames per second using 2/3 pulldown, progressive scan mode. At the time of writing, the only video camera which actually does this is the Panasonic ag100 and is only available as a 24P camera in NTSC. It is perhaps a touch ironic that after all the technological changes seen in the last few years the digital form of storytelling has returned to its original roots which is to mimic film.

You can rest assured, if you ever do get the opportunity to edit a production destined for theatrical release, then Final Cut Pro and Cinema Tools are standing by. Your post-production budget will be such that you will own the equipment you cut with.

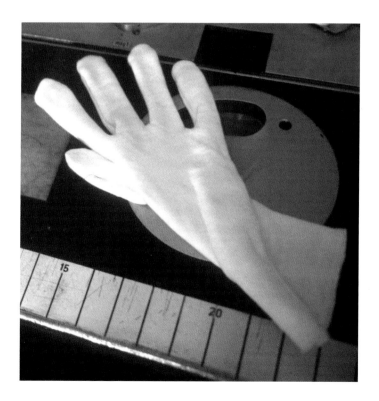

End.

Index

Adding tracks, 70–1, 170–1
Animation, 133–4, 140, 143, 193
Apple computers, 4–5, 207, 214
Applications folder, 11
Arrange options, 21–4
Arrow tool, 71, 76–8
Aspect ratio, 9
Audio, 159–72
   adding tracks, 170–1
   compression, 182
   Cross Fades, 169–70
   Digital Video, 13–14
   dissolves, 170
   effects, 87
   fades, 164–6, 169–70
   Flow direction, 67–9
   keyframes, 167–9
   levels, 160–1, 163–4, 169
   Linked Selection, 74
   Meters, 21–3, 160, 167
   Mixdown, 171–2, 174
   Mixer, 166–8
   rendering, 104
   sample rates, 13–14, 43–4
   sound mix, 161, 164
   Soundtrack, 199–211
   Split Edits, 90–3
   sync, 82–3
   Transitions, 169
Auto-Select Toggle facility, 87

Backups, 174, 176
Batch Capture, 33, 35, 38–41
Beats Per Minute (BPM), 209
Betacam, 7
Bins, 49–52
BPM see Beats Per Minute
Break-off tabs, 67–8, 93
Browsers, 21, 85, 190–2
Button bars, 26, 28

Cables, Firewire, 6–7, 47
Canvas, 21, 61, 65, 139–40
Capture, 2, 29–44

Capture Clip, 33, 35, 36–7
Capture Now, 33, 35, 37–8
Capture Preset, 14–16, 39
CD-ROM, 176, 179, 180
CDs see Compact discs
Character generators, 184
Cinema Tools, 213
Clip Overlays, 25, 133, 156–7, 163
Clips:
   bins, 50–1
   capture, 33, 36–7, 38–40, 41
   Drag and Drop Editing, 93–6
   duration, 35, 146
   effects, 115–16, 151–2
   extending, 96–7
   Linked Selection, 74
   Match Frame Editing, 86
   Multi-Layered Dissolves, 142
   offline, 108–9
   Razorblade tool, 79
   reducing, 96–7
   searching, 53–4
   size, 125
   slow/fast motion, 88–90
   speed, 149–51
   Start/Stop Detection, 48
   Stereo Pairs, 161–3
   Subclips, 84–5
   sync, 82–3
   Timeline edit, 75–6
   viewing, 46, 51–2
Close Gap, 75–6
Color bars, 102–4, 175
Compact discs (CDs), 42–3
   see also CD-ROM
Compositing:
   definition, 3
   LiveType, 189–90
   video effects, 115–16, 122–3
   Viewer Setup, 147
Compression, 176–82
Connectors, Firewire, 6–7
Conversion:
   audio sample rates, 43–4

INDEX

Conversion (*Continued*)
   QuickTime, 180–2
   Stereo Pairs, 161–3
Copying, 77–8, 151–2
   see also Capture
Crop facility, 132
Cross Fades, 169–70
Custom Layouts, 23–4, 27–8, 134
Cut, 78
Cutoff, Titles, 157
Cuts, 141

Deck control, 34
Delete key, 76
Deletion:
   Arrow tool, 77
   bins, 51
   clips, 51, 108–9
   keyframes, 139
   Render Manager, 111
   tracks, 70–1
Digital laboratory, 2–3
Digital Video (DV), 8
   audio, 13–14, 160–72
   capture methods, 33–41
   DV 32 kHz, 14–17
   DV-NTSC, 9, 13–14, 85
   DV-PAL, 9, 13–14, 85
   effects (DVE), 122–4, 131
   hardware requirements, 4
   Start/Stop Detection, 47–9
Dissolves, 119, 141–3, 170
Distribution formats, 176
Drag and Drop Editing, 93–6
Dragging:
   audio tracks, 170
   clips, 96–7
   images, 129, 136
   keyboard shortcuts, 67
   keyframes, 139
Drop Shadow, 127, 133, 156
Dropped frames, 175
Dubbing, 14
Duration:
   clips, 35, 146
   dissolve, 119
   edits, 64
   effects, 139
   Transitions, 118–19

DV *see* Digital Video
DVDs, 176–8, 180
DVE *see* Digital video effects

Easy Setup, 12–13, 14–17
Editing, 55–98
   clips, 53–4
   Drag and Drop, 93–6
   initial process, 2, 57–61
   moving edits, 75–6
   options, 65–6
   ordering material, 49
   output, 174–6
   Split Edits, 90–3
   Timeline, 75–8
   titles, 184
   tools, 71–2
   traditional process, 47–8, 65
Effects:
   audio, 87
   duration, 139
   Emagic, 211
   LiveType, 190, 192, 193–5
   moving, 193–5
   rendering, 100–1, 109–10
   text, 184
   video, 87, 113–57
EMA *see* Essential Message Area
Emagic effects, 211
Essential Message Area (EMA), 130
Expanding sequences, 63
Exporting:
   compression, 177–80
   mixing, 210–11
   QuickTime conversion, 180–2
   Soundtrack, 202–4
   tracks, 210
Extending clips, 96–7

Fades, 141–3, 164–6, 169–70, 209
Fast motion, 88–90, 146, 149–50
Filmmaking process, 56
Filters, 86, 115–17, 119–22, 146
Final Cut Pro:
   benefits, 214–15
   exporting for Soundtrack, 202–4
Fine control, 97
Firewire, 6–9, 13, 47, 60
Fisheye filter, 146

216

Fit to Fill, 66
Fit To Window, 25–6
Flow direction, 67–9
Focus, 146
Fonts, 191–2, 195–6
Footage:
 capture, 33–41, 42
 organization, 45–54
Formats:
 capture, 32–3
 video, 9
Frames:
 dropped, 175
 Freeze Frame, 85–6
 Match Frame Editing, 86–8
 Poster Frames, 52–3
Freeze Frame, 85–6

Gaps, 75–6, 96–7
Gaussian Blur, 144
Generators:
 character, 184
 effects, 115–17
 Matte, 156–7
 Text, 153, 155–6
 Video, 153

Handles, 117
Hard drive, 5–6, 31–2, 110
Hardware requirements, 4, 100–1, 200
High definition format, 33

Icon View, 51–2, 94
Image + Wireframe, 128–9, 130
Images see Video
Importing:
 music, 42–3
 Soundtrack, 204–6
In points:
 capture, 35, 37
 Drag and Drop Editing, 95–6
 editing, 58–9
 modifying, 66–7
 rendering, 105
 Split Edits, 90–1
 Three Point Editing, 64–5
Initial Setup, 11–12
Insert Edit, 56–7, 59, 61–4
 audio tracks, 170
 cut, copy, paste, 78
 Drag and Drop, 94–5
 locking tracks, 70
 Timeline, 75–7
Inspector, LiveType, 192
Installation, software, 10–11
Interface, 19–28
Internet, 176, 179–81
Interviews, 90

Jogging clips, 46
Join Through Edit, 80–1

Keyboard:
 layout, 27–8
 shortcuts, 66–7
Keyframing, 133–41
 audio, 167–9
 fades, 164–5
 Filters, 143–6
 Multi-Layered Dissolves, 143
 Time Remapping, 146–51

Labels, clips, 53–4
Laboratory, 2–3
Layering:
 effects, 114–15
 keyframing, 134–5
 Multi-layered effects, 131–3, 141–3
 text, 156
 tracks, 122
Levels, audio, 160–1, 163–4, 169
Linked Selection, 73–4, 80
List view, 51–2
LiveType, 3, 183–97
 compositing, 189–90
 Inspector, 192
 installation, 10
 manual, 193–6
 output, 196–7
 templates, 185–9
Locked tracks, 69–70, 74, 92
Logging clips, 39
Loops, music, 200–1, 207, 210

Mac:
 components, 5
 speakers, 207
 see also Apple computers
Magnifier tool, 72, 79, 81–2, 92

217

INDEX

Make Offline, 108–9
Manual LiveType, 193–6
Masters, output, 176
Match Frame Editing, 86–8
Matte Generator, 156–7
Media:
   Browser, 190–2
   Limit, 92–3, 97, 117
   management, 107–12, 203
Meters, audio, 160, 167
Mixdown Audio, 171–2, 174
Mixing:
   exporting, 210–11
   sound, 2, 161, 164, 166–8
   Soundtrack, 204, 206–10
Monitoring buttons, 24–5
Motion tab, 124–8, 134, 136–9, 150
MPEG-2, 176–9
Multi-layered effects, 114–15, 131–3, 141–3, 197
Multiple:
   clips, 50
   items, 76–8
   Sequences, 83
   tracks, 123–4
Music, 42–3, 200–1, 206–8, 210

Names:
   bins, 49–50
   clips, 37–9, 53–4
   Sequences, 83–4
Now button, 38
NTSC, 9, 13–14, 85

Objects, LiveType, 191–2
Offline:
   clips, 108–9
   items, 40–1
Opacity, 133, 142, 156
Operating system, 5
Optical printers, 122–3
Out points:
   capture, 35, 37
   Drag and Drop Editing, 95–6
   editing, 58–9
   modifying, 66–7
   rendering, 105
   Split Edits, 90–1
   Three Point Editing, 64–5

Output, 3, 173–82, 196–7
Overlays:
   Clip, 25, 133, 156–7, 163
   keyframes, 167
   Titlesafe, 130
Overscan, 130
Overwrite Edit, 56–7, 59, 61–4
   audio tracks, 170
   cut, copy, paste, 78
   Drag and Drop, 94–5
   locking tracks, 70
   Split Edits, 91
   Timeline, 75–7
   Titles, 153

PAL, 9, 13–14, 85
Paste, 78, 151–2
Patch facility, 67
Pen tool, 72, 142–3, 164–5
Picture in picture, 126, 131
Pointer tool, 71, 75
Positioning:
   edits, 64
   images, 128–32
   text, 155–6
Poster Frames, 52–3
Print to Video, 174–6
Processes, 2–3
Pull focus, 146
Pumping, 160

QuickTime, 10–11, 176, 180–2

Radio buttons, 167
Ramping, 146–51
Razorblade tool, 72, 79–81, 92
Real-time (RT) Effects:
   audio keyframes, 167
   Filters, 120–1
   hardware requirements, 4, 200
   rendering, 100–1
   Transitions, 118
Record monitor, 20–1, 65, 86
Recording:
   audio keyframes, 167–9
   sound, 160, 200
Redo, 72

Reduction:
    Canvas, 139–40
    clips, 96–7, 125
Render Manager, 109–11
Rendering, 89–90, 99–106
    LiveType, 196
    media management, 109–11
    Print to Video, 174
    settings, 101–5
    templates, 189
Replace Editing, 66
Reverse play, 90, 150
Rotating images, 126–7, 129, 134, 136–8
Royalties, 200
RT *see* Real-time Effects
Rubberbanding, 164, 166

Safe RT, 101
Sample rates:
    audio, 13–14
    conversion, 43–4
Scale, 124–5, 137
Scene Detection, 48
Scratch disks, 30–2, 108
Scrubber Bar, 46, 137, 139, 154
Searching, clips, 53–4, 109
Selection:
    capture, 40–1
    Linked, 73–4
    multiple items, 76–8
    tracks, 71, 78
Sequences:
    Drag and Drop Editing, 93
    Insert Edit, 57, 60, 62–3
    LiveType, 196
    Match Frame Editing, 86
    new, 83–4
    Overwrite Edit, 57, 60, 62–3
    rendering, 102, 105
    Soundtrack, 202
Setup, 11–13
Shadows, 190
Shortcuts, 66–7
Shots, 79
Single layer effects, 114–15
Size:
    clips, 125
    images, 138

Timeline, 165–6
    tracks, 204
Skipping, 61, 79
Slider bars, 62, 204
Slow motion, 88–90, 146, 149–50, 151
Snapping, 79, 83, 204
Software:
    installation, 10–11
    requirements, 4, 100–1
Sony, 7–8
Sound:
    fades, 164–6
    mixing, 2, 161, 164
    recording, 160
    see also Audio
Soundtrack, 2, 199–211
    exporting, 202–4
    hardware requirements, 200
    importing, 205–6
    installation, 10
    workflow, 204–11
Source monitor, 20–1, 65, 86
Speaker icon, 24–5
Speakers, 207
Speed:
    Capture Clip, 36–7
    clips, 149–51
    slow/fast motion, 88–90
    Time Remapping, 146–51
Split Edits, 90–3
Stacking *see* Layering
Start/Stop Detection, 47–9
Stereo Pairs, 161–3, 168
Still Images, 182
Storage:
    bins, 51
    hard drive space, 6
    scratch disks, 32
Subclips, 84–5
Superimpose, 66, 143
Sync:
    clips, 82–3
    drift, 14

Televisions:
    aspect ratio, 9
    Titlesafe, 130–1
    video, 47
Templates, 185–9

INDEX

Text:
  effects, 184
  Generator, 153, 155–6
  layering, 156
  LiveType, 190, 193–6
  positioning, 155–6
  templates, 187–8
  Titles, 152–7
Textures, LiveType, 191–2, 193–4, 196
Three Point Editing, 64–5
Three-way dissolve, 142
Time Code Display, 35
Time Remapping, 146–51
Timeline, 21–3
  Audio Mixer, 167
  audio tracks, 170–1
  Canvas, 61, 65
  Drag and Drop Editing, 93–5
  expanding, 81
  Insert Edit, 63
  moving edits, 75–6
  multiple item selection, 76–8
  Overwrite Edit, 63
  red bar, 103–4
  render settings, 101–2
  size, 165–6
  slider bar, 62
  Soundtrack, 203–4
  speaker icon, 24–5
  Time Remapping, 149
Titles, 152–7, 184
Titlesafe, 130–1
Tools, 21–3
  Arrow, 71, 76–8
  Cinema Tools, 213
  editing, 71–2
  Magnifier, 72, 79, 81–2, 92
  Pen, 72, 142–3, 164–5
  Pointer, 71, 75
  Razorblade, 72, 79–81, 92
Tracks:
  adding, 70–1
  audio, 170–1
  Audio Mixer, 166–7
  Audio/Video flow direction, 68–9
  deleting, 70–1
  effects, 114
  exporting, 210

  layering, 122
  Linked Selection, 74
  locked, 69–70, 74, 92
  MultipleTracks, 123–4
  selection, 71
  size, 204
  Stereo Pairs, 161
  Timeline edit, 77–8
  volume, 209
Transitions:
  applications, 117–18
  audio, 169
  duration, 118–19
  video effects, 115–16, 122
Trim bins, 49

U-matic, 8
Uncompressed format, 33
Underscan, 130
Undo, 72
Unlimited RT, 101
Unlinked selection, 73–4

Video:
  compositing, 122–3
  compression, 182
  copying effects, 151–2
  effects, 87, 113–57
  Filters, 119–22
  Firewire, 47, 60
  Flow direction, 67–9
  formats, 9
  Generators, 153
  Image + Wireframe, 128–9
  keyframing, 133–41, 143–6
  Linked Selection, 74
  LiveType, 196
  Motion tab, 124–8
  multi-layered effects, 114–15, 131–3, 141–3
  MultipleTracks, 123–4
  printing, 174–6
  rendering, 104
  Split Edits, 90–3
  sync, 82–3
  templates, 188–9
  text, 193–4

Time Remapping, 146–51
Titles, 152–7
Titlesafe, 130–1
Transitions, 117–19, 122
Viewer, 21
   compositing Setup, 147
   Soundtrack, 203
   Three Point Editing, 65
Viewing clips, 46, 51–2

Virtual VTR Controller, 35
Volume, tracks, 209
VTR Controller, 35
VX-1000 (Sony), 8

Widescreen anamorphic format, 9
Window layout, 24
Windows, Capture, 34–6
Workflow, Soundtrack, 204–11